Springer Series in
MATERIALS SCIENCE 93

Springer Series in
MATERIALS SCIENCE

Editors: R. Hull   R. M. Osgood, Jr.   J. Parisi   H. Warlimont

The Springer Series in Materials Science covers the complete spectrum of materials physics, including fundamental principles, physical properties, materials theory and design. Recognizing the increasing importance of materials science in future device technologies, the book titles in this series reflect the state-of-the-art in understanding and controlling the structure and properties of all important classes of materials.

76 **Spirally Anisotropic Composites**
By G.E. Freger, V.N. Kestelman, and D.G. Freger

77 **Impurities Confined in Quantum Structures**
By P.O. Holtz and Q.X. Zhao

78 **Macromolecular Nanostructured Materials**
Editors: N. Ueyama and A. Harada

79 **Magnetism and Structure in Functional Materials**
Editors: A. Planes, L. Manõsa, and A. Saxena

80 **Ion Implantation and Synthesis of Materials**
By M. Nastasi and J.W. Mayer

81 **Metallopolymer Nanocomposites**
By A.D. Pomogailo and V.N. Kestelman

82 **Plastics for Corrosion Inhibition**
By V.A. Goldade, L.S. Pinchuk, A.V. Makarevich and V.N. Kestelman

83 **Spectroscopic Properties of Rare Earths in Optical Materials**
Editors: G. Liu and B. Jacquier

84 **Hartree–Fock–Slater Method for Materials Science**
The DV–X Alpha Method for Design and Characterization of Materials
Editors: H. Adachi, T. Mukoyama, and J. Kawai

85 **Lifetime Spectroscopy**
A Method of Defect Characterization in Silicon for Photovoltaic Applications
By S. Rein

86 **Wide-Gap Chalcopyrites**
Editors: S. Siebentritt and U. Rau

87 **Micro- and Nanostructured Glasses**
By D. Hülsenberg, A. Harnisch, and A. Bismarck

88 **Introduction to Wave Scattering, Localization and Mesoscopic Phenomena**
By P. Sheng

89 **Magnetoscience**
Magnetic Field Effects on Materials: Fundamentals and Applications
Editors: M. Yamaguchi and Y. Tanimoto

90 **Internal Friction in Metallic Materials**
A Reference Book
By M.S. Blanter

91 **Time-dependent Mechanical Properties of Solid Bodies**
A Theoretical Approach
By W. Gräfe

92 **Solder Joint Technology**
Materials, Properties, and Reliability
By K.N. Tu

93 **Materials for Tomorrow**
Theory, Experiments and Modelling
Editors: S. Gemming, M. Schreiber, and J-B. Suck

S. Gemming · M. Schreiber · J.-B. Suck
(Eds.)

Materials for Tomorrow

Theory, Experiments and Modelling

With 92 Figures, 7 in Colour

 Springer

PD Dr. Sibylle Gemming
Institute of Ion Beam Physics
and Materials Research
Forschungszentrum Dresden-Rossendorf e.V.
PF 51 01 19
D-01314 Dresden, Germany
E-Mail: s.gemming@fzd.de

Prof. Dr. Jens-Boie Suck
Institute of Physics
University of Technology Chemnitz
Reichenhainer Straße 70
D-09107 Chemnitz, Germany
E-mail: suck@physik.tu-chemnitz.de

Prof. Dr. Michael Schreiber
Institute of Physics
University of Technology Chemnitz
Reichenhainer Straße 70
D-09107 Chemnitz, Germany
E-mail: schreiber@physik.tu-chemnitz.de

Series Editors:

Professor Robert Hull
University of Virginia
Dept. of Materials Science and Engineering
Thornton Hall
Charlottesville, VA 22903-2442, USA

Professor Jürgen Parisi
Universität Oldenburg, Fachbereich Physik
Abt. Energie- und Halbleiterforschung
Carl-von-Ossietzky-Strasse 9–11
26129 Oldenburg, Germany

Professor R. M. Osgood, Jr.
Microelectronics Science Laboratory
Department of Electrical Engineering
Columbia University
Seeley W. Mudd Building
New York, NY 10027, USA

Professor Hans Warlimont
Institut für Festkörper-
und Werkstofforschung,
Helmholtzstrasse 20
01069 Dresden, Germany

Library of Congress Control Number: 2006935255

ISBN 978-3-540-47970-3 Springer Berlin Heidelberg New York

Springer is a part of Springer Science+Business Media.
springer.com

© Springer-Verlag Berlin Heidelberg 2007

Typesetting: Digital data supplied by editors
Production: LE-TEX Jelonek, Schmidt & Vöckler GbR, Heidelberg, Germany
Cover production: WMX Design GmbH, Heidelberg

Printed on acid-free paper   SPIN: 11874423   57/3100/YL   5 4 3 2 1 0

Preface

Materials science has assumed a key position for new technological developments, and is therefore strongly supported by industry and governments. Nowadays it occupies a bridging position between physics, chemistry and engineering and extends from basic science in physics and chemistry on the atomic scale to large-scale applications in industry. The increasing number of materials science study courses at universities underlines the present and future importance of understanding and developing materials for the future. The contributions to this book evolved from the lectures of such a course, namely the Heraeus summer school on "New Materials for Today, Tomorrow and Beyond", held at Chemnitz University of Technology in October 2004.

Looking at the rapidly developing communications industry, it is obvious that a large part of current research for future materials is devoted to the understanding of materials to be used in nanometer (nm) scale devices. It is therefore not surprising that five of the six lectures collected in this volume are devoted more or less directly to the nm scale, while the first one treats one of the central unsolved problems of condensed matter physics and – because of the widespread industrial application of glasses in every day life – touching also materials science. Some of the materials discussed here are already used today, for others the development has proceeded far enough that their application can be expected in the near future. Yet others, however, we will not be ready to use until some years from now, assuming that the difficulties connected with their application are solved in the near future.

The above-mentioned first article, written by one of the best known scientists in the field of computer simulations, introduces the molecular dynamics (MD) methods used in the calculation of the structure and atomic-scale dynamics of new materials. The method has now gained an admirable predictive power in materials science, explaining details not or hardly accessible by experiments. This has become possible through ab initio calculations of the electronic structure of the material using density functional theory (DFT), which provide very realistic results. These can then be used to model the force field, in which the atoms interact, for MD simulation of larger systems.

In this way undercooled silicate-based fluids and glasses have been successfully investigated including the as yet poorly understood transition from the fluid to the glassy phase.

The same kind of technique was also used to investigate inorganic nanotubes. Carbon nanotubes (NT) have been at the centre of interest since their discovery in 1991, followed by organic NT $B_xC_yN_z$. Both of them are already used in many applications and will find more in the future. However, since then dozens of other inorganic materials have been discovered, which form NT as well. The starting materials are layered compounds, characterized by strong covalent bonds within the layers and van-der-Waals forces between them. Their structure is often complicated and their properties not well known, even though they have great application potential. Simulations of inorganic NT are therefore presented in the second contribution, with a view to better understanding the properties of these special NT.

Nanotechnology also plays a decisive role in information technology. However, the rapid increase (doubling of the Internet traffic every 6 months, of the wireless capacities every 9 months and of the magnetic information storage every 15 months) cannot be compensated by a corresponding shrinkage of the semiconductor devices, as it was achieved in the past 30 years. To keep up with the demands, completely new devices have to be invented, operating on the nanoscale and exploiting quantum effects. One possibility is to use the spin of the electron in addition to its charge for information transmission and storage, i.e. going from conventional electronics to spintronics. The foundations of this technique, exploiting the giant magnetoresistance and the tunneling magnetoresistance are discussed from the experimental and theoretical point of view in the third contribution.

Interfaces play a very important role for the stability and functioning of materials, sometimes as barriers, sometimes mediating between adjacent grains or layers, sometimes protecting the material next to them, avoiding overheating or corrosion in everyday materials. Homophase (between two materials of the same chemical composition and atomic structure) grain boundaries (GB) and heterophase (between two chemically or structurally different materials) interfaces are discussed from a theoretical point of view in the fourth contribution. Electronic and atomic structures at planar GB and heterophase interfaces are investigated and classified according to the dominant interaction determining their properties.

For many technical applications, the dimensions of semiconductor devices are rapidly approaching length scales at which a classical description of their electronic structure and transport properties is insufficient and quantum effects must be taken into account. But since these devices are still too large for an atomistic description, continuum models with empirically adjusted material parameters are successfully used to describe their properties. These continuum models and their implementation in a simulation software are described in the fifth contribution.

When working with devices on the nanoscale one necessarily runs into the problem that some physically significant lengths, like a correlation length, a screening length, the mean free path, etc. start to exceed the dimension of the devices. This necessarily leads to a drastic change of the physical properties in comparison to those in the same material at larger dimensions. Such a change of properties with respect to the melting temperature, magnetic and mechanical properties and finally to the atomic dynamics is discussed at the example of metallic nanocrystals in the final contribution to this book.

As the lectures were given to students of different disciplines (physics, chemistry, materials science), the articles describe not only our current knowledge in each of the fields, but also basic facts needed for their understanding. Thus the articles combine basic information at a textbook level with a presentation of cutting-edge research in the field and an outlook to future developments, thereby bridging the gap between specialized reviews and study books. In addition, lectures on materials simulation, a very rapidly advancing discipline, are combined with lectures on experimental research in materials science, and the techniques used in both of these disciplines.

The editors of this book are indebted to the lecturers of the summer school for taking on the burden of writing up their contributions in addition to presenting them. Finally we would like to thank the WE Heraeus Stiftung for enabling us to organize this school by its generous financial support.

Chemnitz, July 2006   *Sibylle Gemming*
Michael Schreiber
Jens-Boie Suck

Contents

1 Computer Simulations of Undercooled Fluids and Glasses
K. Binder, D. Herzbach, J. Horbach, M.H. Müser 1
1.1   Introduction 1
1.2   A Tutorial Review of the MD Technique 4
       1.2.1   The Verlet Algorithm 4
       1.2.2   How to Estimate Intensive Thermodynamic Variables
                from Microcanonical MD Runs;
                Realization of Other Ensembles 7
       1.2.3   Diffusion, Hydrodynamic Slowing Down, Einstein
                and Green–Kubo Relations 9
1.3   A Comparative Test of Model Potentials for Silica 12
1.4   Simulations of Molten and Glassy Silicon Dioxide 23
1.5   Mixtures of Silicon Dioxide with Sodium Oxide
       and Aluminium Oxide...................................... 27
1.6   Conclusions 29
References 30

2 Simulation of Inorganic Nanotubes
A.N. Enyashin, S. Gemming, G. Seifert 33
2.1   Introduction 33
2.2   Design of Inorganic Nanotubes 34
2.3   General Criteria for the Stability of Inorganic Nanotubes 38
2.4   Theoretical Prediction of the Properties
       of Non-Carbon Nanotubes 40
       2.4.1   Nanotubes of the IVA Group Elements 41
       2.4.2   Nanotubes of the VA Group Elements 43
       2.4.3   Nanotubes of Boron and Borides 43
       2.4.4   Nanotubes of Boron Nitride and its Analogues 45

2.4.5   Nanotubes of Chalcogenides 47
2.4.6   Nanotubes of Oxides 50
2.5   Conclusion.. 54
References 55

**3 Spintronics: Transport Phenomena
in Magnetic Nanostructures**
P. Zahn 59
3.1   Introduction 59
3.2   Magnetism in Nanostructures 61
      3.2.1   Magnetism in Reduced Dimensions 61
      3.2.2   First Principals Calculational Scheme 62
      3.2.3   Magnetic Interlayer Exchange Coupling 64
3.3   Transport Phenomena.................................. 66
      3.3.1   Transport Theory 66
              Diffusive and Coherent Transport Regime............... 67
              Boltzmann Theory................................. 68
              Residual Resistivity 69
              Landauer Theory 70
      3.3.2   Giant Magnetoresistance 72
              Basics... 72
              Microscopic Origin................................ 73
              Applications 77
      3.3.3   Tunneling Magnetoresistance........................ 78
              Basics... 78
              Microscopic Origin................................ 80
              Applications 85
References 86

4 Theoretical Investigation of Interfaces
S. Gemming, M. Schreiber 91
4.1   Interfaces – Boundaries Between Two Phases.................. 91
      4.1.1   Introduction 92
      4.1.2   Interactions 95
              Coulomb Interaction 95
              Elastic Interaction 96
              Electron Transfer 97
              Pauli Repulsion 98
              Image Charge Interaction 98
4.2   Theoretical Methods 98
      4.2.1   Theory of the Electronic Structure..................... 99
              Density-Functional Theory........................... 99
              Computational Details of Bulk State Calculations 100
              Density-Functional Tight-Binding Approaches 100

      4.2.2   Classical Modelling 101

              Image-Charge Models 101

              Effective Many-Body Potentials 102

              Ionic Models 103

4.3   Homophase Boundaries.................................... 104

      4.3.1   Pristine Boundaries 105

      4.3.2   Non-Stoichiometric and Doped Boundaries.............. 108

4.4   Heterophase Boundaries 109

      4.4.1   Wetting and Growth of Metal Layers.................. 110

      4.4.2   Metal–Ceramic Boundaries 112

      4.4.3   Reactive Metal–Semiconductor Interfaces 116

4.5   Summary and Outlook 118

References 119

5 Electronic Structure and Transport for Nanoscale Device Simulation

A. Trellakis, P. Vogl.. 123

5.1   Introduction 123

5.2   Electronic Structure of Semiconductors 125

      5.2.1   Bloch Theory and the Band Structure.................. 125

      5.2.2   The k·p-Approximation 127

      5.2.3   Conduction and Valence Band Models................. 128

5.3   Heterostructures... 130

      5.3.1   The Envelope Function Approximation 130

      5.3.2   Elastic Deformation and Strain 131

      5.3.3   Carrier Densities at Non-Zero Temperature 132

      5.3.4   Charge Distributions and the Poisson Equation......... 134

5.4   Carrier Transport in Nanostructures 135

      5.4.1   Classical Ballistic Transport 135

      5.4.2   Scattering and the Boltzmann Equation 136

      5.4.3   The Drift-Diffusion Equations 138

      5.4.4   Quantum Corrected Drift-Diffusion 139

      5.4.5   Quantum Ballistic Transport 140

5.5   The nextnano³ Simulation Package 141

      5.5.1   Capabilities Overview 141

      5.5.2   Numerical Methods 142

      5.5.3   Example Application................................. 144

References 145

6 Metallic Nanocrystals and Their Dynamical Properties

Jens-Boie Suck 147

6.1   Introduction 147

6.2   Production of Nanocrystalline Materials 150

6.3   Characterization of Nanocrystalline Materials 155

6.4   Some General Properties of Grains
        and Grain Boundaries of Nanocrystals..........................158
6.5   Some Examples of the Special Properties
        of Nanocrystals...160
        6.5.1   Melting Temperature 160
        6.5.2   Some Magnetic Properties
                of Nanocrystalline Materials 161
        6.5.3   Some Mechanical Properties
                of Nanocrystalline Materials 164
6.6   Vibrational Properties
        of Metallic Nanocrystalline Materials...........................168
        6.6.1   General Remarks on the Atomic Dynamics
                of Metallic Nanocrystals 168
        6.6.2   Phonon Confinement...................................171
        6.6.3   Grain-Size Dependence of the Atomic Dynamics 172
        6.6.4   Comparison with the Atomic Dynamics
                of Metallic Nanocrystals and the Related
                Amorphous Solids 175
        6.6.5   Specific Investigations Concerning the Contribution
                of the Grain Boundaries and Surfaces
                to the Observed Spectra..............................176
References 184

Index 191

List of Contributors

Kurt Binder
Institut für Physik
Johannes Gutenberg-Universität
Mainz
Staudinger Weg 7
D-55099 Mainz
Germany
kurt.binder@uni-mainz.de

Andrey N. Enyashin
Institute of Solid State Chemistry
Ekaterinburg 620219
Russia
Institute of Physical Chemistry
and Electrochemistry
D-01062 Dresden
Germany
Enyashin@ihim.uran.ru

Sibylle Gemming
Institute of Ion Beam Physics
and Materials Research
FZ Rossendorf
D-01314 Dresden
Germany
Institute of Physical Chemistry
and Electrochemistry
D-01062 Dresden
Germany
s.gemming@fz-rossendorf.de

Daniel Herzbach
Institut für Physik
Johannes Gutenberg-Universität
Mainz
Staudinger Weg 7
D-55099 Mainz
Germany
herzbach@uni-mainz.de

Jürgen Horbach
Institut für Physik
Johannes Gutenberg-Universität
Mainz
Staudinger Weg 7
D-55099 Mainz
Germany
horbach@uni-mainz.de

Martin H. Müser
Department of Applied
Mathematics
University of Western Ontario
London, Ontario N6A5B7
Canada
mmuser@uwo.ca

Michael Schreiber
Institute of Physics
University of Technology Chemnitz
D-09107 Chemnitz
Germany
schreiber@physik.tu-chemnitz.de

Gotthard Seifert
Institute of Physical Chemistry
and Electrochemistry
D-01062 Dresden
Germany
gotthard.seifert
  @chemie.tu-dresden.de

Jens-Boie Suck
Institute of Physics
University of Technology Chemnitz
D-09107 Chemnitz
Germany
suck@physik.tu-chemnitz.de

Alex Trellakis
Walter-Schottky Institute
Technische Universität München
D-85748 Garching
Germany
alex.trellakis@wsi.tum.de

Peter Vogl
Walter-Schottky Institute
Technische Universität München
D-85748 Garching
Germany
vogl@wsi.tu-muenchen.de

Peter Zahn
Fachbereich Physik
Martin-Luther-Universität
Halle-Wittenberg
D-06099 Halle
Germany
peter.zahn@physik.uni-halle.de

1

Computer Simulations of Undercooled Fluids and Glasses

Kurt Binder, Daniel Herzbach, Jürgen Horbach, and Martin H. Müser

Abstract. An introduction to the Molecular Dynamics (MD) simulation of chemically realistic models for undercooled fluids and glasses is given, emphasizing silicatic materials such as molten silicon dioxide and its mixtures with sodium oxide and aluminium oxide, and comparing the simulation results to experimental data whenever possible.

A key ingredient to the computer simulation of materials is a sufficiently accurate description of the force fields with which the atoms interact. The need to simulate large systems for sufficiently long times makes the use of effective potentials for classical MD methods desirable. The validation of such effective potentials is best done studying the corresponding crystalline states of the material. As an example, the use of the so-called BKS-potential studying the structural and thermal properties of quartz crystals is described, and a comparison to other potentials is discussed.

When one studies undercooled fluids and glasses, a second problem enters, the disparity between experimental cooling rates and the much larger rates of the simulation. The extent to which a meaningful comparison to experiments is nevertheless possible is discussed. It is shown that the simulations can reproduce the structural and dynamic properties of molten silica (including self diffusion coefficients, viscosity, sound velocity, etc.) and its mixtures with other oxides. Evidence for the formation of sodium-rich channels responsible for anomalously large diffusion constants of sodium in mixtures containing sodium oxide will be discussed. No enhanced diffusion of aluminium occurs, however, due to "tricluster" formation in such mixtures.

1.1 Introduction

In this section first the scope and purpose of Molecular Dynamics Simulations will be reviewed, considering also the relation of this method to other computer simulation techniques briefly. Then the basic facts about MD algorithms are reviewed and some simulation "knowhow" will be mentioned.

All computer simulations of materials aim to calculate their structure and dynamics, using some atomistic input [1–7]. Depending on the purpose of the modelling, this input can differ very much. Here we emphasize at the outset that we are neither concerned with very small scale phenomena (1 Å or less), where the focus is on electrons and properties related to the electronic structure [5], nor with very large scale phenomena (100 Å or more), as one needs to consider particularly in the context of "soft materials" (polymers, colloidal suspensions, microemulsions, etc.) [6, 7]. All these problems require special techniques different from the methods suitable for the simulation of the structure and dynamics on the scale between 1 Å and 100 Å, on which we focus here.

The theoretical basis in physics for all methods of computer simulation of materials is statistical thermodynamics and quantum mechanics. The conceptually simplest approach is classical MD [1–3], however, where one simply solves numerically Newton's equations of motion for the interacting many-particle system (atoms or molecules interacting with effective potentials). This method thus is based on classical mechanics: one creates a deterministic trajectory in the phase space of the system. One then takes time averages of the observables of interest along this trajectory, relying on the ergodicity hypothesis of statistical mechanics, which says that these time averages are equivalent to ensemble averages in the microcanonical (NVE) ensemble. This type of simulation utilizes a constant number of particles (N) in a fixed volume (V) and Newton's equations of motion conserve the total (internal) energy E, of course. Already at this point one sees that the conjugate intensive thermodynamic variables (temperature T, pressure p) can only be inferred indirectly and will exhibit statistical fluctuations: unlike experiments, the finite particle number N is not astronomically large, and hence such fluctuations typically are not negligible and need careful consideration: we shall come back to this issue in Sect. 1.2.

Sometimes it offers advantages to realize other ensembles of statistical mechanics via a MD simulation directly, such as the constant volume-constant temperature (NVT) ensemble or the NpT ensemble. This is possible by introducing a coupling to appropriate "thermostats" or "barostats" (Sect. 1.2). Alternatively, T can be held constant by introducing a weak friction force, together with random forces, whose strengths are controlled by the fluctuation-dissipation theorem. Such techniques are routinely used for simulating polymer melts [8, 9]. This method is closely related to stochastic simulation methods such as "Brownian Dynamics", where one simulates a Langevin equation (if the friction is strong enough, the inertial term in the equation of motion can be omitted). Of course, dynamical correlation functions obtained from such methods differ somewhat from those obtained from strictly microcanonical MD. But for the computation of static properties along the stochastic trajectory in phase space such methods can be very advantageous.

This latter statement also holds for the importance sampling Monte Carlo (MC) method [2–4, 10]. There one generates a random-walk like trajectory in phase space, controlled by transition probabilities that ensure the approach to thermal equilibrium via the so-called "detailed balance condition". Many of the practical aspects of computer simulations, such as "statistical errors" and systematic errors due to the finite size of the simulated system or the finite "length" of the trajectory that is generated (which means that the observation time is finite, and often even fairly small, see Sect. 1.2) are shared by all these simulation methods.

MC methods have the advantage in general that the "MC move" effecting the transition from one microstate to the next one along the stochastic trajectory is rather arbitrary, and hence can be adapted to the problem at hand. Also one can realize rather straightforwardly any ensemble of statistical mechanics that is useful for the considered problem. E.g., when one wants to study properties along the coexistence curve where the A-rich phase of a binary (A, B) mixture of Lennard-Jones particles can exist in equilibrium with the B-rich phase, it is convenient to use the "semi-grandcanonical ensemble" where the chemical potential difference $\mu_A - \mu_B$ (= 0, if the mixture is symmetrical) is the given independent variable (in addition to T, V, and $N = N_A + N_B$). MC moves then are constructed where an A-particle transforms into B or vice versa [11,12] (the same method is also useful for solid binary alloys [13]). Using such well-equilibrated states as initial states for strictly microcanonical MD runs, one can study dynamical correlations and transport coefficients precisely at the coexistence curve [12], avoiding any problems with artefacts due to thermostats.

While hence MC, as is well known, can have many advantages for the simulation of fluids [1–4], there are also cases where really efficient MC moves are not yet known, and MD is even preferable for equilibration. This is the case for the systems of interest here, molten silicon dioxide and its mixtures with other oxides, because of the presence of rather strong covalent "chemical bonds" in this material. Therefore we shall not describe the technical aspects of MC methods here.

So far we have ignored the basic fact that the basic physics of condensed matter is described by quantum mechanics rather than classical mechanics. However, attempting a numerical solution of the Schrödinger equation for a system of many nuclei and electrons is still premature and not yet feasible even on the fastest computers. Thus, one has to resort to approximations. A very successful approach is the "ab initio MD" or "Car Parrinello Molecular Dynamics" (CPMD) [14], where the degrees of freedom of the electrons are included via density functional theory (DFT) [15]. The huge advantage of this technique is that one no longer relies on effective interatomic potentials. The latter often are phenomenological and chosen ad hoc, lacking any firm foundation from quantum chemistry [5]. But CPMD also suffers from a serious disadvantage: it is several orders of magnitude slower than classical MD, and hence only very short timescales and very small systems are accessible. Moreover, the method has problems to treat van der Waals-type forces between neutral atoms, such as occur in gases like argon or krypton, where simple Lennard–Jones (LJ) potentials (perhaps amended with three-body forces) will work better. Also, usually the degrees of freedom of the nuclei (or ions, respectively) are still treated classically.

A complementary approach uses again effective potentials between ions and/or neutral atoms, but takes the quantum effects for ionic motion into account via Path Integral Monte Carlo (PIMC) [16–20] or Path Integral Molecular Dynamics (PIMD) [21–24]. For a description of the thermal properties of solids at low temperatures (specific heat, thermal expansion, etc.) use of such techniques indeed is indispensable to ensure that these properties comply with the third law of thermodynamics. For most fluids (quantum liquids such as ^{4}He which can become superfluid and ^{3}He need to be treated by PIMC [17,18], of course) classical MD is good enough, and hence is in the focus of the remainder of these lecture notes. However, there are problems like proton transport in liquid water, where quantum effects of the protons need to be included as well as electronic degrees of freedom (formation and breaking of hydrogen bonds). Then a combination of CPMD and PIMD is needed to handle such a problem [25]. However, we shall not describe such recent advances here, and only recall the basic facts about classical MD in the next section. In the third section, we shall discuss the proper choice of suitable effective potentials, using silicon dioxide as an example. We shall show that the simulation of structural and thermal properties of the various crystalline phases of silica provide a stringent test to evaluate the quality of potentials. In the fourth section, we shall review the studies of molten and glassy silica, paying attention to the problem of equilibration in view of the very fast cooling implemented by MD. Sect. 1.5 discusses the extension of such work to mixtures of silicon dioxide with sodium oxide and aluminium oxide, while Sect. 1.6 summarizes our conclusions.

1.2 A Tutorial Review of the MD Technique

1.2.1 The Verlet Algorithm

The simplest case to consider is a system of N atoms with Cartesian coordinates $\boldsymbol{X} = \{\boldsymbol{r}_i\}$, $i = 1, \ldots, N$, in d dimensions. According to classical mechanics, the dynamics is described by Newton's equations of motion $\{\dot{\boldsymbol{r}}_i \equiv d\boldsymbol{r}_i/dt = \boldsymbol{v}_i$, the velocity$\}$,

$$m_i \ddot{\boldsymbol{r}}_i = -\frac{\partial U_{\text{pot}}}{\partial \boldsymbol{r}_i} = \boldsymbol{f}_i \, , \tag{1.1}$$

m_i being the mass of the i'th particle and $\boldsymbol{f}_i$ the force acting on it. We assume this force is only due to interactions with other particles, described by a potential $U_{\text{pot}}(\boldsymbol{X})$,

$$U_{\text{pot}} = \sum_{i=1}^{N-1} \sum_{j>i}^{N} U(\boldsymbol{r}_{ij}), \qquad \boldsymbol{r}_{ij} = \boldsymbol{r}_i - \boldsymbol{r}_j \, . \tag{1.2}$$

For simplicity, in Eq. (1.2) it is assumed that U_{pot} is pairwise additive, and hence

$$\boldsymbol{f}_i = -\sum_{j(\neq i)} \frac{\partial U(r_{ij})}{\partial \boldsymbol{r}_j} = \sum_{j(\neq i)} \boldsymbol{f}_{ij} \,. \tag{1.3}$$

The total energy E is conserved,

$$E = E_{\text{kin}} + U_{\text{pot}} = \sum_{i=1}^{N} \frac{1}{2} m_i \dot{\boldsymbol{r}}_i^2 + U_{\text{pot}} \,, \tag{1.4}$$

$$\frac{dE}{dt} = \sum_{i=1}^{N} m_i \dot{\boldsymbol{r}}_i \ddot{\boldsymbol{r}}_i - \sum_{i=1}^{N} \dot{\boldsymbol{r}}_i \cdot \boldsymbol{f}_i = 0 \,. \tag{1.5}$$

In a MD simulation Newton's equations of motion are integrated numerically. A computationally efficient scheme to do this is the Verlet algorithm [26], for instance. It can be derived by expanding $\boldsymbol{r}_i(t \pm \delta t)$ in a Taylor series with respect to the time increment δt,

$$\boldsymbol{r}_i(t+\delta t) = \boldsymbol{r}_i(t) + \delta t\, \boldsymbol{v}_i(t) + \frac{1}{2m_i}(\delta t)^2 \boldsymbol{f}_i(t) + \frac{1}{6}(\delta t)^3 \boldsymbol{b}_i(t) + o\left((\delta t)^4\right) \,, \tag{1.6}$$

$$\boldsymbol{r}_i(t-\delta t) = \boldsymbol{r}_i(t) - \delta t\, \boldsymbol{v}_i(t) + \frac{1}{2m_i}(\delta t)^2 \boldsymbol{f}_i(t) - \frac{1}{6}(\delta t)^3 \boldsymbol{b}_i(t) + o\left((\delta t)^4\right) \,, \tag{1.7}$$

$\boldsymbol{b}_i(t)$ denoting the vector appearing in the third order of the expansion. For $\ddot{\boldsymbol{r}}_i$ we have already substituted Eq. (1.1). Adding Eqs. (1.6) and (1.7), the odd orders cancel, and thus

$$\boldsymbol{r}_i(t+\delta t) = 2\boldsymbol{r}_i(t) - \boldsymbol{r}_i(t-\delta t) + \frac{1}{m_i}(\delta t)^2 \boldsymbol{f}_i(t) + o\left((\delta t)^4\right) \,. \tag{1.8}$$

Subtraction of Eq. (1.7) from Eq. (1.6) yields an equation for the velocity

$$\boldsymbol{v}_i(t) = \frac{1}{2(\delta t)} [\boldsymbol{r}_i(t+\delta t) - \boldsymbol{r}_i(t-\delta t)] + o\left((\delta t)^3\right) \,. \tag{1.9}$$

This algorithm [1–3,26] is time-reversible: interchanging $\boldsymbol{r}_i(t+\delta t)$ and $\boldsymbol{r}_i(t-\delta t)$ yields the propagator for the time evolution to go backwards in time, and the sign of $\boldsymbol{v}_i(t)$ is changed.

Note that the updating of the velocities is one step behind, $\boldsymbol{v}_i(t)$ can only be calculated after $\boldsymbol{r}_i(t + \delta t)$ has become available. Positions and velocities are updated in synchronized fashion by the "Velocity Verlet" algorithm [27]. Using

$$\boldsymbol{r}_i(t+\delta t) = \boldsymbol{r}_i(t) + \delta t\, \boldsymbol{v}_i(t) + \frac{1}{2m_i}(\delta t)^2 \boldsymbol{f}_i(t) \tag{1.10}$$

and the corresponding time-reversed equation

$$\boldsymbol{r}_i(t) = \boldsymbol{r}_i(t+\delta t) - \delta t\, \boldsymbol{v}_i(t+\delta t) + \frac{1}{2m_i}(\delta t)^2 \boldsymbol{f}_i(t+\delta t) \tag{1.11}$$

one finds

$$v_i(t + \delta t) = v_i(t) + \frac{\delta t}{2m_i}[f_i(t) + f_i(t + \delta t)]. \qquad (1.12)$$

The updating of the velocities now requires that the forces of both the present and the next configuration are known (which also need the coordinates at time $t + \delta t$). The Verlet algorithm and the Velocity Verlet algorithm yield identical trajectories, both algorithms are completely equivalent to each other.

It is of interest to clarify how large the time step should be. Consider liquid argon, as an example: the atoms are assumed to interact with a Lennard–Jones (LJ) potential,

$$U(r_{ij}) = 4\epsilon \left[\left(\frac{\sigma}{r_{ij}} \right)^{12} - \left(\frac{\sigma}{r_{ij}} \right)^6 \right], \qquad (1.13)$$

with $\sigma \approx 3.4 \, \text{Å}$, $\epsilon/k_B \approx 120 \, \text{K}$. Rescaling coordinates as $r^* = r/\sigma$, one gets

$$r^*(t + \delta t) = 2r^*(t) - r^*(t - \delta t) - \left(\frac{\delta t}{\tau_0} \right)^2 \frac{r_{ij}^*}{|r_{ij}^*|} \sum_{j(\neq i)} \left[|r_{ij}^*|^{-13} - \frac{1}{2}|r_{ij}^*|^{-7} \right], \qquad (1.14)$$

where we have introduced the natural time unit τ_0 of MD as

$$\tau_0 = \left(\frac{m\sigma^2}{48\epsilon} \right)^{1/2}, \qquad (1.15)$$

which becomes roughly $\tau_0 \approx 3.1 \times 10^{-13} \, \text{s}$, noting the mass $m \approx 6.6 \times 10^{-23} \, \text{g}$ for argon atoms. In order to keep numerical integration errors small, we need $\delta t/\tau_0 \ll 1$, e.g. $\delta t^* = 0.03$, i.e. a time step $\delta t \approx 10^{-14} \, \text{s}$. Thus even a million MD time steps ($t^* = t/\tau_0 = 30000$) only corresponds to a very short physical time, of the order of $10 \, \text{ns}$.

It is rather instructive (and a useful starting point for the discussion of more advanced algorithms) to go beyond this simple approach and to discuss the time evolution in phase space more formally in terms of the Liouville operator $\hat{\mathcal{L}}$. One can formally write for any observable A

$$A(\Gamma, t) = \hat{U}(t)A(\Gamma, 0), \qquad (1.16)$$

$\Gamma \equiv (X, P)$ where X combines all the Cartesian coordinates of the N particles and P their momenta, $P = (p_1, p_2, \ldots, p_N)$, into a point in a $2Nd$-dimensional space, d being the spatial dimensionality. The propagator $\hat{U}(t)$ is a unitary operator and can be written as

$$\hat{U}(t) = \exp(i\hat{\mathcal{L}}t). \qquad (1.17)$$

An important aspect of this description is Liouville's theorem, which states that the flow in phase space is incompressible, i.e. phase space volume is conserved. The Verlet algorithm has the property that unitarity is preserved to order $(\delta t)^3$, and it is a particularly stable algorithm over rather long times. Tuckerman *et al.* [28] give a more extensive discussion on the stability and other general properties of MD algorithms.

1.2.2 How to Estimate Intensive Thermodynamic Variables from Microcanonical MD Runs; Realization of Other Ensembles

If the algorithm conserves the energy E rigorously, one would realize states distributed according to the microcanonical (NVE) ensemble of statistical mechanics. Due to integration errors, the energy is not strictly conserved, however; but if the time step is chosen small enough, this problem may be disregarded. Assuming that the microcanonical ensemble is realized perfectly, how can we estimate the thermodynamic variables of interest, such as the temperature T and pressure p?

One important observation is that in classical statistical mechanics the distributions of the positions of the particles and their momenta factorize; since one has a simple Maxwell–Boltzmann distribution for the velocities of the particles, T can be inferred from their kinetic energy. We define the "kinetic temperature" $\mathcal{T}$ by (for $d = 3$)

$$\mathcal{T} = \frac{2}{3k_B N} E_{\text{kin}} = \frac{1}{3k_B N} \sum_{i=1}^{N} m_i v_i^2, \qquad (1.18)$$

and the desired estimate of the system temperature then is $T = \langle \mathcal{T} \rangle$. Of course, for finite N there occur temperature fluctuations in the NVE ensemble [29],

$$\frac{[\langle \mathcal{T}^2 \rangle - \langle \mathcal{T} \rangle^2]}{\langle \mathcal{T} \rangle^2} = \frac{2}{3N} \left(1 - \frac{3k_B}{2C_V} \right), \qquad (1.19)$$

with C_V being the specific heat of the system, at constant volume. Note that the estimation of temperature using $T = \mathcal{T}$ involves a statistical error, of course, since the MD run is carried out over a finite observation time only.

The estimation of the pressure is possible with the help of the virial theorem. The dynamical variable $\mathcal{P}$

$$\mathcal{P} = \frac{1}{3V} \left(2E_{\text{kin}} + \sum_{i=1}^{3N} \boldsymbol{r}_i \cdot \boldsymbol{f}_i \right), \qquad (1.20)$$

when averaged over yields the hydrostatic pressure p,

$$p = \langle \mathcal{P} \rangle = \frac{N k_B T}{V} + \frac{1}{3V} \sum_{i=1}^{N} \langle \boldsymbol{r}_i \cdot \boldsymbol{f}_i \rangle. \qquad (1.21)$$

The first term in this expression occurs already in the ideal gas, the second term represents the effect of the intermolecular interactions.

Of course, due to the fact that the time step δt is not infinitesimally small the algorithm defined in Eqs. (1.8) and (1.9) does not conserve the energy strictly. In early MD work researchers therefore rescaled the velocity from time to time, to correct for the drift in the energy. Note however, that then one no longer realizes strictly the microcanonical ensemble (or any other standard ensemble) of statistical mechanics, and also the time evolution of the system (and hence time-dependent correlation functions that one would like to extract) get somewhat disturbed. The latter problem is also true for an alternative method, where from time to time one reassigns the velocities by randomly drawing them from a Maxwell–Boltzmann distribution.

In statistical mechanics for the computation of averages of static quantities, one often would prefer to realize the canonical (NVT) ensemble, where the absolute temperature T is a given variable and therefore does not fluctuate at all. It then is the total energy E which fluctuates in a finite system, of course, and again these fluctuations can be related to the specific heat C_V, via the fluctuation relation

$$\frac{C_V}{k_B} = \frac{1}{N(k_B T)^2} \left[\langle \mathcal{H}^2 \rangle_{NVT} - \langle \mathcal{H} \rangle^2_{NVT} \right] , \qquad (1.22)$$

$\mathcal{H}$ being the Hamiltonian of the system. If one uses the method to rescale the velocities, then both E and T fluctuate, and neither of the fluctuation relations, Eq. (1.19) or Eq. (1.22), holds.

However, one can formulate a MD algorithm that yields precisely the desired (NVT) ensemble, if one extends the Lagrangian in order to include the coupling to a thermostat. This means that one adds a friction-like term to Newtonian equations of motion,

$$\ddot{\boldsymbol{r}}_i = \frac{\boldsymbol{f}_i}{m_i} - \xi \dot{\boldsymbol{r}}_i , \qquad (1.23)$$

and the "friction coefficient" $\xi(t)$ itself fluctuates in time around zero, according to the equation of motion

$$\dot{\xi} = \frac{1}{Q} \left[\sum_{i=1}^{N} m_i v_i^2 - 3N k_B T \right] , \qquad (1.24)$$

the parameter Q being a (fictitious) mass of the thermostat. In this Nosé–Hoover algorithm [30, 31] $\xi(t)$ reacts to the imbalance between the instantaneous kinetic energy and its (intended) canonical average (remember $\sum_i m_i \langle v_i^2 \rangle = 3N k_B T$). Of course, for Eqs. (1.23) and (1.24) the total energy is no longer strictly conserved; rather one can identify a different conserved energy-like quantity $\mathcal{H}'$,

$$\mathcal{H}' = \mathcal{H} + \frac{1}{2} Q \xi^2 + 3N k_B T \int \xi(t') dt' . \qquad (1.25)$$

Also in this case it is true that the dynamical time-displaced correlation functions are not precisely the same as the microcanonical ones, although in favorable cases the differences are negligibly small [32]. However, if one wishes to combine strictly Newtonian equations of motion with a canonical ensemble, one can realize this by a hybrid MC–MD approach, generating an ensemble of well-equilibrated configurations by canonical NVT Monte Carlo runs, which are used as initial states for microcanonical MD runs [12].

If one wishes to directly realize the isothermal-isobaric (NpT) ensemble, where the pressure is given and the volume is a fluctuating variable, one can do this by coupling the system to the so-called "Andersen barostat" [33].

1.2.3 Diffusion, Hydrodynamic Slowing Down, Einstein and Green–Kubo Relations

In this subsection we discuss how one estimates the various transport coefficients from MD simulations, and related technical aspects.

The phenomenological description of diffusion can be conveniently based on Fick's law, where one considers a gradient in the density $\rho(\boldsymbol{r}, t)$ of particles, and the resulting current density $\boldsymbol{j}(\boldsymbol{r}, t)$ that acts to reduce the gradient,

$$\boldsymbol{j}(\boldsymbol{r}, t) = -D\nabla\rho(\boldsymbol{r}, t)\,, \tag{1.26}$$

D being the diffusion constant. Combining this (phenomenological) constitutive equation of irreversible thermodynamics with the (exact) continuity equation, which expresses the conservation law for the total number N of the particles,

$$\frac{\partial\rho(\boldsymbol{r}, t)}{\partial t} + \nabla \cdot \boldsymbol{j}(\boldsymbol{r}, t) = 0\,, \tag{1.27}$$

yields the diffusion equation,

$$\frac{\partial\rho(\boldsymbol{r}, t)}{\partial t} = D\nabla^2\rho(\boldsymbol{r}, t)\,. \tag{1.28}$$

Introducing Fourier components $\delta\rho_{\boldsymbol{k}}(t)$ of the density fluctuations

$$\delta\rho(\boldsymbol{r}, t) \equiv \rho(\boldsymbol{r}, t) - \langle\rho\rangle = \int \delta\rho_{\boldsymbol{k}}(t)\,\exp(i\boldsymbol{k}\cdot\boldsymbol{r})\,d\boldsymbol{k}\,, \tag{1.29}$$

the diffusion equation is readily solved, yielding a simple exponential relaxation,

$$\frac{d}{dt}\delta\rho_{\boldsymbol{k}}(t) = -Dk^2\delta\rho_{\boldsymbol{k}}(t), \quad \delta\rho_{\boldsymbol{k}}(t) = \delta\rho_{\boldsymbol{k}}(0)\exp\left(-Dk^2 t\right)\,. \tag{1.30}$$

From the argument of the exponential in Eq. (1.30) one can read off the associated relaxation time $\tau_{\boldsymbol{k}}$,

$$\tau_{\boldsymbol{k}} = \left(Dk^2\right)^{-1}\,. \tag{1.31}$$

It is seen that this relaxation time $\tau_{\boldsymbol{k}}$ diverges for $k \to 0$, and hence the dynamic correlation function of density fluctuations in a diffusive system decays very slowly for long wavelengths,

$$S(\boldsymbol{k}, t) \equiv \langle\delta\rho_{-\boldsymbol{k}}(0)\delta\rho_{\boldsymbol{k}}(t)\rangle = S(\boldsymbol{k})\exp\left(-Dk^2 t\right)\,, \tag{1.32}$$

where $S(\boldsymbol{k})$ is the static structure factor,

$$S(\boldsymbol{k}) = \langle \delta\rho_{-\boldsymbol{k}}(0)\delta\rho_{\boldsymbol{k}}(0)\rangle \,. \tag{1.33}$$

Equations (1.31) and (1.32) exemplify the well-known "hydrodynamic slowing down" [34], the slow decay of long wave length density fluctuations being caused by the conservation law for this quantity. On the one hand, the analysis of the intermediate scattering function $S(\boldsymbol{k}, t)$ as defined in Eq. (1.32) allows to estimate the diffusion constant D as

$$D = \lim_{k\to 0} \left\{ \frac{1}{k^2 t} \ln\left[\frac{S(\boldsymbol{k}, t)}{S(\boldsymbol{k})}\right]\right\} \tag{1.34}$$

for large enough times t. On the other hand, hydrodynamic slowing down makes the equilibration of simulations in the NVT ensemble difficult, at least for large systems. This difficulty applies to both MD and MC simulations. Remember that for a simulation of fluids, it is often convenient to use as an initial configuration a crystal, putting the atoms on the regular positions of a lattice that fits in the simulation box. Using a cubic box with linear dimensions L and applying periodic boundary conditions, the smallest wave number is $k_{\min} = 2\pi/L$ and hence the largest relaxation time is $\tau_{\max} = (Dk_{\min}^2)^{-1} = L^2/(4\pi^2 D)$. In undercooled fluids the self diffusion constant D of the particles is rather small, and hence $\tau_{\max}$ is very large. In particular for large systems the question to what extent full thermal equilibrium has been reached requires careful consideration.

We now consider the solution of the diffusion equation, Eq. (1.28), in real space, considering $\rho(\boldsymbol{r}, t)$ as the density of a (single) labelled particle in the fluid, such that $\int \rho(\boldsymbol{r}, t)d\boldsymbol{r} = 1$. Choosing as an initial condition a delta function, $\rho(\boldsymbol{r}, t = 0) = \delta(\boldsymbol{r} - \boldsymbol{r}_0)$, $\rho(\boldsymbol{r}, t)$ is nothing but the van Hove self correlation function, that describes the conditional probability of finding a particle at $\boldsymbol{r}$ after a time t provided this particle was at $\boldsymbol{r} = \boldsymbol{r}_0$ at time $t = 0$ [35]. The corresponding solution of Eq. (1.28) then is simply a Gaussian distribution,

$$\rho(\boldsymbol{r}, t) = \frac{1}{(4\pi Dt)^{d/2}} \exp\left[-\frac{(\boldsymbol{r} - \boldsymbol{r}_o)^2}{4Dt}\right], \tag{1.35}$$

where d denotes the dimensionality of space (usually $d = 3$). The squared half width of this distribution increases linearly with time, and so does the mean square displacement

$$\langle [\delta\boldsymbol{r}(t)]^2\rangle = \langle [\boldsymbol{r}(t) - \boldsymbol{r}_0]^2\rangle = 2dDt, \quad t \to \infty\,. \tag{1.36}$$

This is the famous Einstein relation. We have added here the restriction to consider large times, $t \to \infty$, since then this equation holds quite generally, while typically the simple diffusion equation will not hold on small length scales (of the order of a few atomic diameters or so) and the associated short times, but only on large scales of both length and time. This fact will be evident from the examples discussed later. In fact, Eq. (1.36) is routinely used to estimate the self diffusion coefficient in simulations. Of course, Eq. (1.36) also applies to more-component systems where one wishes to distinguish between the diffusion coefficients D_α of the different kinds of particles α.

In order to provide a motivation for the existence of Green–Kubo relations between transport coefficients and suitable integrals of time-dependent correlation functions, we sketch the derivation of the relation between the self diffusion constant and the velocity autocorrelation function of a diffusing particle [35]. Writing

$$\boldsymbol{r}(t) - \boldsymbol{r}_0 = \int_0^t \boldsymbol{v}(t')dt' \,, \tag{1.37}$$

the mean square displacement becomes

$$\langle[\boldsymbol{r}(t) - \boldsymbol{r}_0]^2\rangle = \int_0^t dt' \int_0^t dt'' \langle \boldsymbol{v}(t'') \cdot \boldsymbol{v}(t')\rangle = d \int_0^t dt' \int_0^t dt'' Z(t'' - t') \,, \tag{1.38}$$

with $Z(t'' - t') \equiv \langle v_\alpha(t'')v_\alpha(t')\rangle$ being a correlation function of one of the d Cartesian components of $\boldsymbol{v}(t)$. Since in equilibrium translational invariance holds with respect to the origin of time, $Z(t'' - t')$ can only depend on the difference of the two times t'', t', but not on them separately. Rearranging the domain of integration in Eq. (1.38) yields

$$\langle[\boldsymbol{r}(t) - \boldsymbol{r}_0]^2\rangle = 2dt \int_0^t (1 - s/t)Z(s)ds \to 2dDt \,, \quad t \to \infty \,, \tag{1.39}$$

where

$$D = \int_0^\infty Z(s)ds = \int_0^\infty \langle v_\alpha(0)v_\alpha(t)\rangle dt \,. \tag{1.40}$$

Thus the self diffusion constant is the time integral of the velocity autocorrelation function.

Equation (1.40) is a special case of a Green–Kubo relation [35]. Similar relations exist for the bulk viscosity η_B and the shear viscosity η [36]

$$\eta_B = \frac{V}{k_B T} \int \langle J_{\alpha\alpha}(t)J_{\alpha\alpha}(0)\rangle \,, \quad \eta = \frac{V}{k_B T} \int_0^\infty dt \langle \sigma_{\alpha\beta}(t)\sigma_{\alpha\beta}(0)\rangle \,, \quad \alpha \neq \beta \,, \tag{1.41}$$

with $J_{\alpha\alpha}(t) = p(t) - \langle p \rangle$, where $p(t)$ is equal to the diagonal elements of the pressure tensor $\underline{\sigma}$,

$$\sigma_{\alpha\beta} = \frac{1}{V} \sum_{i=1}^N \left[m_i v_{i\alpha}v_{i\beta} + \frac{1}{2} \sum_{j(\neq i)} (\boldsymbol{r}_{ij})_\alpha f_\beta(\boldsymbol{r}_{ij}) \right] \,. \tag{1.42}$$

Similarly one finds for the thermal conductivity λ_T that

$$\lambda_T = \frac{1}{V k_B T^2} \int\limits_0^\infty dt \langle j_z^e(0) j_z^e(t) \rangle , \tag{1.43}$$

with

$$j_z^e = \frac{d}{dt} \left[\sum_{i=1}^N z_i \left(\frac{1}{2} m_i \boldsymbol{v}_i^2 + \sum_{j(\neq i)} U(r_{ij}) \right) \right] . \tag{1.44}$$

1.3 A Comparative Test of Model Potentials for Silica

If one wishes to use simulations in order to make predictions of the properties of specific materials, it is clear that the use of potentials that are accurate enough is crucial. As an example, we shall discuss this problem for fluid and solid silicon dioxide, emphasizing the use of the potential proposed by van Beest, Kramer and van Santen ("BKS potential") [37]

$$U(r_{ij}) = \frac{q_i q_j e^2}{r_{ij}} + A_{ij} \exp(-B_{ij} r_{ij}) - \frac{C_{ij}}{r_{ij}^6} . \tag{1.45}$$

The first term is a Coulomb-like interaction, but with effective charges $q_i \{i \in \text{Si}, \text{O}\}$ rather than the true ionic charges, namely $q_O = -1.2$, $q_{Si} = +2.4$, e being the elementary charge. The constants A_{ij}, B_{ij} and C_{ij} of the short-range Buckingham potential can be found in the original reference [37]. The same form of the potential, but with different parameters, has earlier been proposed by Tsuneyuki et al. [38] (TTAM potential), based on "ab initio" Hartree–Fock selfconsistent field calculations. Van Beest et al. [37] reparametrized this potential, combining quantum chemical "ab initio" computations with fits to some macroscopic experimental data.

It is somewhat surprising that by a clever choice of these phenomenological constants in Eq. (1.45) it is possible to describe the directional covalent bonding. Typically a silicon atom is located in the center of a tetrahedron, with the oxygens at the corners. Although one just uses pair potentials which depend on the absolute value of the distance between atoms only, this directional bonding results from the competition between the Si–O and O–O potentials. Of course, one cannot expect that such a description is perfectly accurate, phenomena such as charge transfer between ions, local fluctuations of charge or of dipole moments etc. cannot be described, since the degrees of freedom of the electrons have already been completely eliminated. Nevertheless, the use of such simple pair potentials is desirable, since such simulations are still technically very difficult: the long range Coulomb interactions necessitates the use of the time-consuming Ewald summation techniques [1–3]; the scale for the potential is in the electronvolt energy range and varies rather

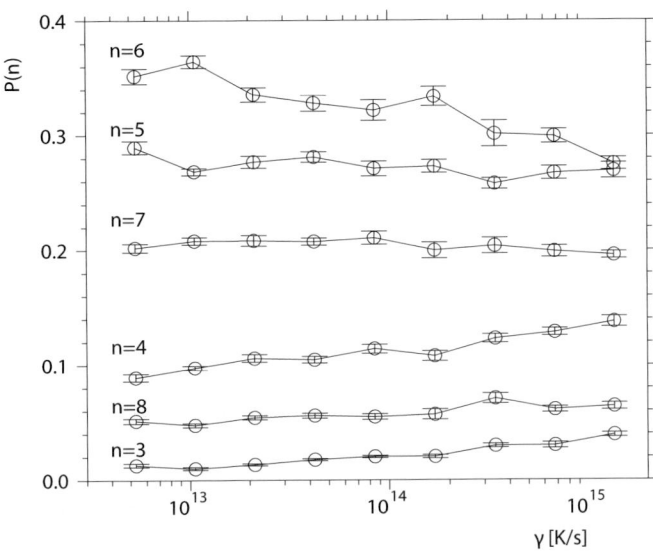

Fig. 1.1. Dependence of the probability $P(n)$ that in the network structure of SiO_2 a ring of length n is present plotted vs. the cooling rate γ. This simulation used 668 oxygen and 334 Si atoms in a cubic box, applying periodic boundary conditions, and cooling the sample at constant pressure in the NpT-ensemble from $T_0 = 7000\,$K to the final temperature $T_f = 0\,$K. An average over 10 independent runs was performed, allowing to estimate the statistical errors as shown in the figure. From Vollmayr *et al.* [39]

rapidly with distance. Therefore one needs to use a rather small MD time step, namely $\delta t = 1.6\,$fs.

In view of the doubts whether such a potential as written in Eq. (1.45) is accurate enough, it is desirable to validate it by a comparison of simulation results with suitable experimental data. Being interested in supercooled fluids and glasses, it is tempting to test the potential for silica Eq. (1.45) by a simulation that yields the properties of silica glass. However, such a test is also problematic, as the following example shows. The most plausible procedure to produce amorphous glassy structures by MD simulations is to imitate the procedure used in the glass factory: One starts at a high temperature T_0 where silica is molten and then one cools the system gradually down at constant pressure, so the temperature $T(t)$ varies linearly with time t, $T(t) = T_0 - \gamma t$, γ being the cooling rate. While the resulting structures look reasonable [39], one must be aware of the fact that the cooling rates applied in the simulation are extremely large ($\gamma = 10^{12}\,$Ks^{-1} or even larger), a factor of 10^{12} (or more) larger than those used in the experiment. Also the initial temperature used in Ref. [39] ($T_0 = 7000\,$K) is by far higher than its experimental counterpart. Thus, one should check to what extent the quantitative details of the physical properties do depend on this cooling history.

As an example we show in Fig. 1.1 how the distribution of the length n of rings in the covalent network structure depends on γ [39]. A ring is defined as the shortest connection of consecutive Si–O bonds that form a closed loop, and n is the number of such bonds. Note that in pure amorphous silicon dioxide at low temperatures the "chemical rule" is strictly obeyed that each silicon atom is connected by bonds to four oxygens, at the corners of a (deformed) tetrahedron, and each such oxygen is shared by two tetrahedra ("bridging oxygen" position) and hence is two fold coordinated. It does neither occur that an oxygen has $z = 3$ nearest silicon neighbors nor $z = 1$ (this would be a "dangling bond" situation), although such values do occur at high temperatures.

It turns out, as discussed in depth in Ref. [39], that many other properties (e.g. the enthalpy, the density, the distribution of the tetrahedral angle θ between two Si–O bonds at a silicon atom, etc.) all depend distinctly on cooling rate. This is not a surprise at all, although such cooling rate effects were denied in very early investigations of glassy materials (e.g. [40]), which did not have the computer resources at their disposal to clearly identify them. It is also obvious from Fig. 1.1, that a unique extrapolation of such data to experimentally relevant cooling rates is somewhat doubtful.

It is clear, however, that the brute force cooling at constant pressure, as used in Ref. [39], is not the most suitable technique to produce a good glassy structure in a simulation. A better method [41–44] is to choose fixed density beforehand (e.g. taken from experiment) and then apply the NVT ensemble, simulating the system at the lowest temperature where it still can be equilibrated with an effort of a few nanoseconds equilibration time. Since one needed to use a box containing 8016 atoms (such a size is just enough to avoid finite size effects [45]), choosing a substantially larger equilibration time would be computationally very costly. The lowest temperature of a SiO_2 melt that thus could be fully equilibrated was $T = 2750$ K, still far above the melting temperature of crystalline SiO_2. Using then such configurations of well equilibrated melts as initial states for a further cooling (applying again cooling rates of the order of $10^{12}\,\mathrm{Ks}^{-1}$ or larger) at constant density, one obtains structures which compare with experiments [46] very well, as far as the neutron structure factor is concerned (Fig. 1.2) [43]. With this type of cooling history and the constant density constraint, one can no longer detect any significant dependence on cooling rate, for the few decades of cooling rates that are accessible in MD simulations. Unfortunately, one can never exclude that for much slower cooling rates (as are used in experiments) still a systematic deviation of the results from those results that are generated with a cooling rate of $10^{12}\,\mathrm{Ks}^{-1}$ could occur.

Therefore, in principle as well as in practice, a more stringent and conclusive test of potentials can be made by checking to what extent the properties of a material in the crystalline state are reproduced by a model potential. Silica has several crystal structures, such as α-quartz and β-quartz, and these structures have been extensively studied by various experimental techniques [47].

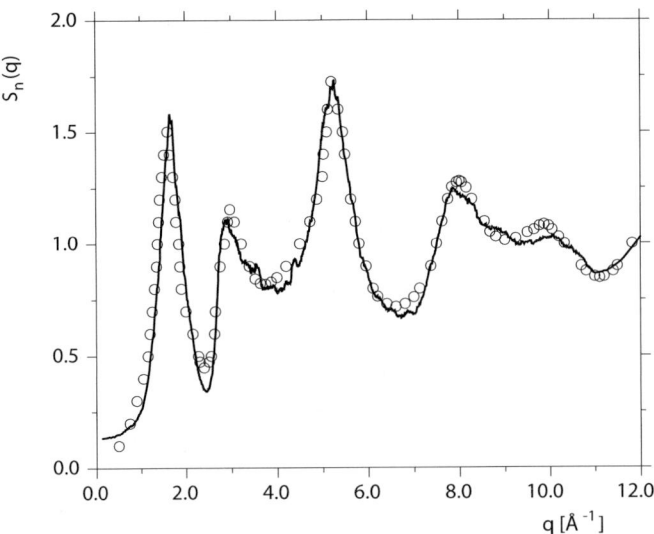

Fig. 1.2. Static neutron structure factor of SiO_2 at room temperature ($T = 300\,\mathrm{K}$), plotted versus wave number q. The full curve is the MD simulation, using the experimental scattering lengths for Si and O atoms, while the symbols are the neutron scattering data of Price and Carpenter [46]. From Horbach and Kob [43]

It then is a sensitive test of the quality of the potential, to what extent the simulations can reproduce the temperature dependence of the various crystal properties, and the location of the phase transition temperatures in the (p, T) phase diagram [48, 49].

For the simulations of crystals at state points sufficiently far away from phase transitions the fact that MD simulations can explore only the nanosecond time range is not a problem, because one can use the information from X-ray diffraction on the crystal structure at least as a good first guess for the initial condition of the structure. Of course, one cannot expect a priori that the lattice parameters of the experiment will be reproduced precisely by any approximate potential. Therefore, it is important to carry out the simulation in a generalization of the NpT-ensemble to anisotropic crystals, the constant stress ensemble [50], and choose a shape of the simulation box (together with the periodic boundary condition) that is compatible with the expected crystal symmetry. In this Parrinello–Rahman method [50], one hence does not work with a cubic simulation box as one does in the simulation of fluids and amorphous solids, but allows for a more general shape that can also fluctuate and get deformed in the course of the simulation, in order to maintain the condition of zero external stress.

Of course, when one works at state points close to structural phase transitions (or close to the melting point), one clearly may run into the standard problems associated with the simulation of all kinds of phase transitions, namely finite observation time effects and finite size effects [4, 10]. At a phase transition which is strongly first order, hysteresis is a problem. At phase transitions which are second order or only weakly first order, one expects a large correlation length ξ of order parameter fluctuations (in such cases one is able to define an order parameter ϕ that distinguishes between these phases in the sense that it is identically zero in one phase and nonzero in the other one, see, e.g., Ref. [51] for a more detailed discussion in the context of materials science). The critical divergence of ξ at second order phase transitions also leads to critical slowing down (the relaxation time τ diverges, $\tau \propto \xi^z$, where z is the "dynamic exponent", with $z \approx 2$ in cases where the order parameter ϕ is not conserved [52]). In a computer simulation, the finite size of the simulation box leads to a finite size rounding of the divergence of both of ξ and of τ, but in any case sufficiently large systems are required (typically $N \geq 10^3$, making the use of "ab initio MD" [5, 14, 15] impossible) and several values of the particle number N need to be used, to check for finite size effects and, if possible, perform a "finite size scaling" [4, 10, 53, 54] analysis of the simulation data.

In the case of the phase transition of SiO_2 from α-quartz to β-quartz, which experimentally is found to be only very weakly of first order [55], a global order parameter of the low-temperature phase (i.e., α-quartz) can be defined that measures the rotation of (distorted) SiO_4-tetrahedra about the [100] axis, such that

$$\phi = \frac{1}{N^*} \sum_{i=1}^{N^*} \varphi_i \,, \tag{1.46}$$

where the sum over i is confined to sites which are equivalent to the sites marked by an arrow in Fig. 1.3, N^* is the total number of such sites, and

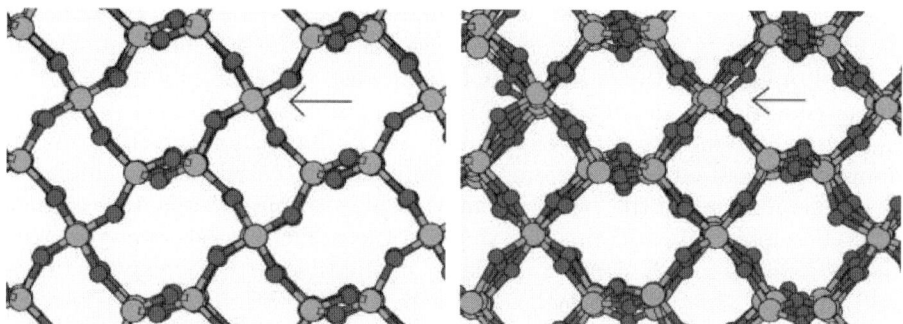

Fig. 1.3. View along the [100] axis in α-quartz at $T = 80\,\mathrm{K}$ (*left*) and β-quartz at $T = 1050\,\mathrm{K}$ (*right*). Dark and light atoms represent oxygen and silicon atoms, respectively. Both snapshots belong to identical subvolumes of the simulation cell. The [001] axis goes from the left to the right. The rotation angles about the [100] axis of the units marked by an arrow are used to define the order parameter (Eq. (1.46)). From Müser and Binder [48]

φ_i denotes the (averaged) deviation of the orientation in the y–z plane of the (four) Si–O bond(s) from the value in the ideal β-quartz structure.

The MD simulations were mostly done with $N = 1080$ atoms, starting from an orthorhombic box with lengths $A = 25.0$ Å, $B = 26.0$ Å and $C = 22.1$ Å. Near the α–β phase transition, also system sizes $N = 2160$ and $N = 4320$ were considered as well, using correspondingly larger boxes [48]. In order to speed up calculations, interactions and forces were tabulated on a grid with a resolution of 5×10^{-4} Å. No significant differences between simulations using the tabulated and the non-tabulated interactions were detected.

Figure 1.4 shows then the temperature variation of the average value of the order parameter. One can see that for $T < 500$ K both sizes shown ($N = 1080$ and 2160, respectively) coincide, so there is no significant finite size effect, but at higher temperatures there are systematic deviations between both sizes, and even in the high temperature phase (β-quartz) $\langle|\phi|\rangle$ is still nonzero (the well-known [10,53,54] "finite size tails" arise because statistical fluctuations of the order parameter ϕ around zero are not yet completely negligible in a finite system and make a nonzero contribution to the absolute value $\langle|\phi|\rangle$). On the basis of the simulation data shown in Fig. 1.4 alone it would not be possible to locate the transition temperature precisely, and to distinguish whether the transition is second order or first order. The transition temperature can

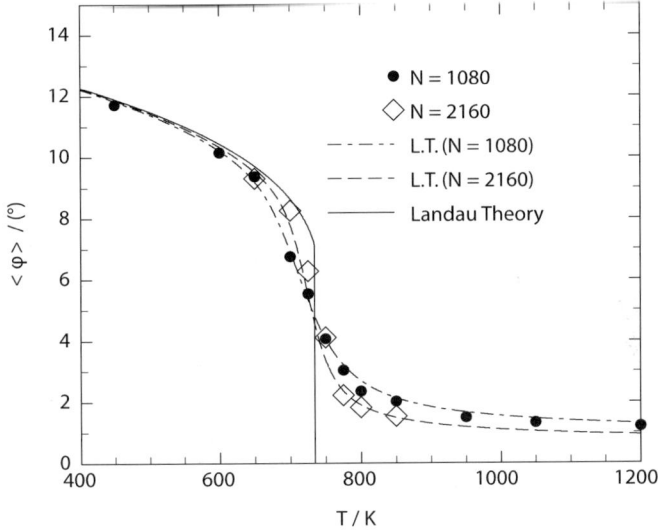

Fig. 1.4. Order parameter $\langle|\phi|\rangle$ as a function of temperature, including simulation results (*full dots* and *open diamonds*) for two system sizes. The lines reflect fits according to Landau theory, where free parameters were adjusted to the $N = 1080$ curve. The solid line corresponds to the thermodynamic limit ($N \to \infty$) in Landau theory (Eq. (1.49)), while broken lines represent finite-size Landau theory (Eq. (1.49)). From Müser and Binder [48]

be located, however, by evaluating the fourth-order cumulant of the order parameter

$$g_4(T) = \frac{1}{2}\left(3 - \frac{\langle \phi^4 \rangle_N}{\langle \phi^2 \rangle_N^2}\right) \tag{1.47}$$

and looking for an intersection point at which several curves of $g_4(T)$ for different N cross [56], see Fig. 1.5 [48]. The resulting estimate for the transition temperature $T_{\text{tr}} = 740 \pm 5\,\text{K}$ deviates from the experimental one by about $100\,\text{K}$, indicating already that the BKS potential is not perfect, as expected, but actually it is quite satisfactory that no larger deviation occurs!

In order to distinguish the order of the transition from simulation methods alone, one also can apply finite size scaling methods [4, 10, 53, 54], but then one needs data over a wide range of N (at least a decade in the variation of N is required). In view of the complicated (long range!) interactions, such a study would require already a substantial effort in computing time resources (the larger N the longer runs are required to equilibrate the system in the transition region, of course [4, 10, 53, 54]). Therefore in Ref. [48] we avoided this effort by using a simple theoretical analysis in terms of a Landau theory, where the free energy $f(\phi, T)$ per atom is expanded in powers of ϕ, in the thermodynamic limit,

$$f(\phi, T) = \frac{1}{2}a(T - T_c)\phi^2 + \frac{1}{4}b\phi^4 + \frac{1}{6}c\phi^6, \tag{1.48}$$

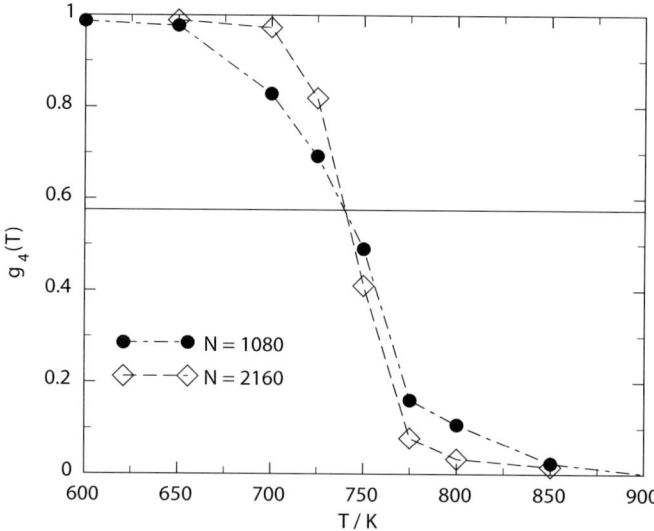

Fig. 1.5. Fourth-order cumulant $g_4(T)$ of the order parameter distribution plotted as a function of temperature T for different system sizes. The value g_4^* at which crossing of the cumulants is predicted within Landau theory is indicated by a straight line. Broken lines are drawn to guide the eye. From Müser and Binder [48]

where a, b, c, and T_c are free parameters. Of course, Eq. (1.48) would not be quantitatively useful for a system with short range forces undergoing a second-order phase transition [53, 57], but it is widely believed (e.g. [55, 58]) that Eq. (1.48) is applicable to study the phase transition from α-quartz to β-quartz. However, in order to find the parameters that are appropriate to describe the simulation results, one needs to generalize the approach to *finite* system sizes. This is done by evaluating numerically expectation values of the moments, $\langle |\phi|^n \rangle$, as follows,

$$\langle |\phi|^n \rangle = \frac{\int\limits_{-\infty}^{+\infty} |\phi|^n \, \exp\left[- \frac{N f(\phi, T)}{k_B T} \right] d\phi}{\int\limits_{-\infty}^{+\infty} d\phi \, \exp\left[- \frac{N f(\phi, T)}{k_B T} \right]} . \tag{1.49}$$

The parameters a, b, c, and T_c ($= 715\,\mathrm{K}$) where determined by fitting $\langle |\phi| \rangle$ for $N = 1080$, and it was very reassuring to find that the same parameters gave a very good description of both $\langle |\phi| \rangle$ and $\langle \phi^2 \rangle$ for all values of N that were studied. In this way, the curves shown in Fig. 1.4 were obtained, and it was hence found that for $N \to \infty$ the transition is weakly of first order ($\langle \phi \rangle$ jumps then discontinuously from about $\langle \phi \rangle \approx 7°$ to zero at T_{tr}, the parameter b in Eq. (1.48) being slightly negative). Qualitatively, though not quantitatively, these findings agree with experiment [55].

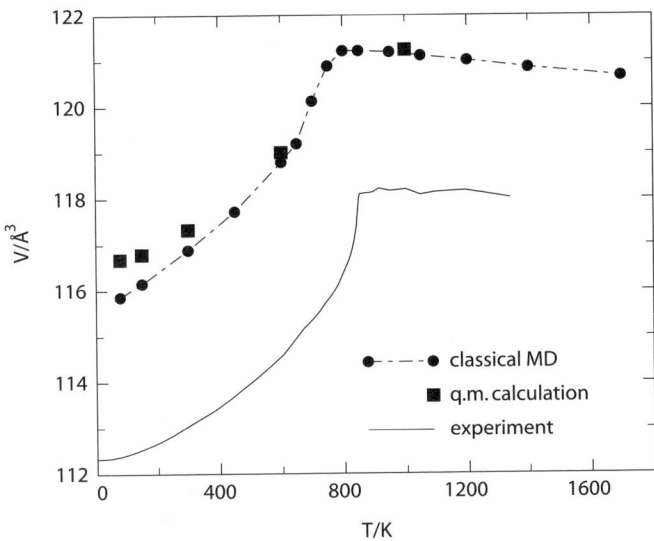

Fig. 1.6. Volume V per unit cell of quartz as a function of temperature T. Classical molecular dynamics simulations, path integral molecular dynamics (refered to as quantum mechanical (q.m.) in the figure) and experimental data [55] are shown. From Müser and Binder [48]

Thus, it is not too surprising that some discrepancies between simulation and experiment are found in other quantities as well, such as the volume V per unit cell, Fig. 1.6, or the c/a ratio, Fig. 1.7 [48]. As a side remark, we mention that a comparison between (classical) MD results for a crystal and experimental data makes only sense at temperatures T above the Debye temperature θ. For SiO_2, $\theta \approx 500$ K [47]. Classical MD predicts, e.g., that the thermal expansion coefficient for $T \to 0$ goes to a (nonzero) constant, and hence at low T the volume increases linearly with temperature, in contradiction with the third law of thermodynamics, which requires the thermal expansion coefficient to vanish as $T \to 0$. Thus for $T \le \theta$ a quantum mechanical simulation method (for the ions or nuclei, respectively) is required, and this can be achieved by the path integral molecular dynamics (PIMD) method [21,23]. In fact, Fig. 1.6 includes such PIMD data for quartz [23], and one sees that the PIMD data for $V(T \to 0)$ approach $V(T = 0)$ with a horizontal slope, as the experimental data do. In fact, if one plots $V(T)/V(T = 0)$ versus T, experiment and PIMD simulation agree quantitatively for $T \le \Theta$ [23].

The relative difference in volume between simulation and experiment is about 3%, in agreement with previous findings [59], while with the TTAM potential [38] the discrepancy is about 7% [60]. More disturbing than this difference is the qualitatively different behavior of the c/a ratio (Fig. 1.7): While the experiment shows a pronounced anomaly at the transition, there is no trace of this anomaly to be seen in the simulation. The absence of this anomaly is not a consequence of finite size rounding, as the comparison

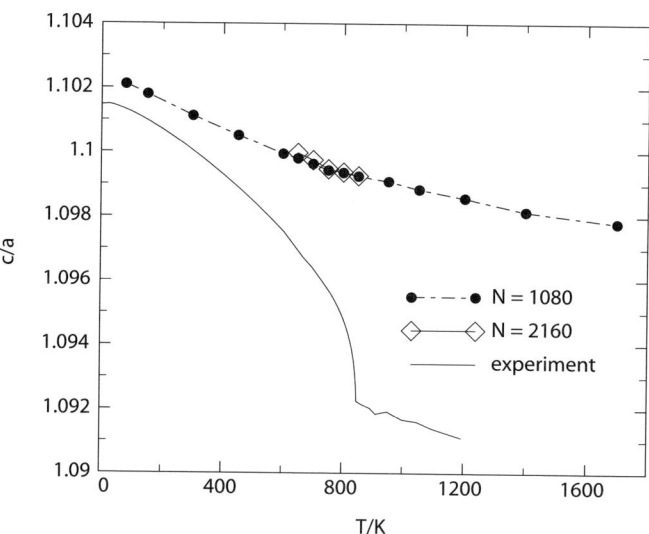

Fig. 1.7. c/a ratio of quartz as a function of temperature T. Two different system sizes are considered. Solid lines correspond to experimental data [55]. From Müser and Binder [48]

between the results for $N = 1080$ and $N = 2160$ shows. This failure to reproduce an anomaly in the c/a-ratio at the $\alpha \rightarrow \beta$ transition of quartz must be attributed to a shortcoming of the potential. Similar problems occur for some of the other crystal structures of SiO_2, too.

In view of the problems, it is of interest to try out more complicated effective potentials for silica and check to what extent the problems with the BKS potential can be remedied. Herzbach *et al.* [49] investigated two potentials: the fluctuating charge potential due to Demiralp, Cagin and Goddard (DCG potential) [61] and the fluctuating dipole moment potential (oxygen atoms being polarizable) due to Tangney and Scandolo (TS potential) [62]. Both potentials (for details see [49,61,62]) need considerably more computational effort (up to two orders of magnitude more computing time is needed). The reason is that fluctuating charge potentials allow the effective ionic charge on an atom to fluctuate as a function of the local environment. Thus, at a given set of coordinates, the charges have to be determined first, before the forces on the ions can be determined. In the TS potential [62], ionic charges are kept fixed, but the additional dipole moments on the oxygen atoms need to be calculated self consistently with the electric field in every MD step.

Figure 1.8 shows a comparison between the predictions for the volume of SiO_2 as a function of temperature, as obtained from these three potentials (the largest system size for which data for all three potentials were available was $N - 864$). It is seen that all three potentials underestimate the

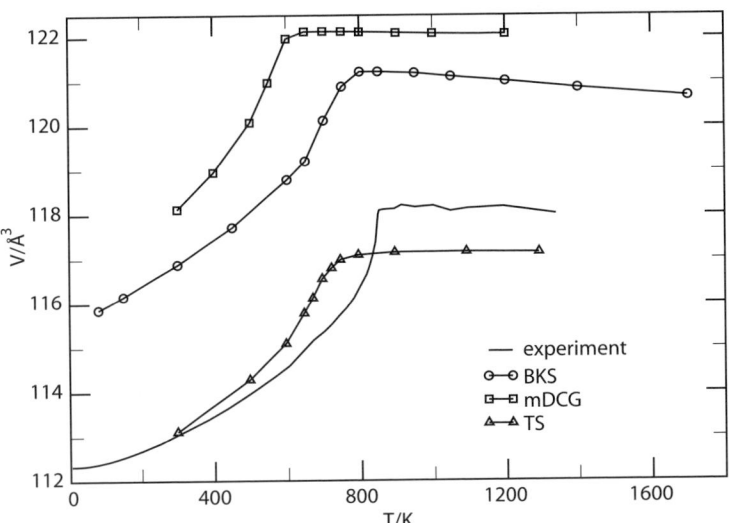

Fig. 1.8. Volume per unit cell of SiO_2 at zero pressure plotted vs. temperature, as compared to experiment (*full curve without symbols*) [35]. Curves with symbols refer to the BKS potential (*circles*), a slightly modified version of the DCG potential (*squares*), and the TS potential (*triangles*). From Herzbach *et al.* [49]

experimental transition temperature of $T_{tr} = 864\,K$, namely $T_{tr}(BKS) = 740\,K$, $T_{tr}(mDCG) = 581\,K$ and $T_{tr}(TS) = 712\,K$. These values were estimated with the cumulant intersection method, as described above. With respect to T_{tr}, the simple BKS performs best! When one compares elastic constants [49, 63], one finds no clear picture, all three potentials deviate from experimental data by similar amounts, so such a comparison does not discriminate between the three potentials. However, when one considers the c/a ratio (Fig. 1.9) then the TS potential has a clear advantage, since it reproduces the anomaly rather well, apart from the shift of T_{tr} and a shift of the ratio c/a at the transition temperature itself. Thus we conclude that among the three tested potentials (BKS, mDCG, TS) the TS potential is the best one, but when one considers all properties of interest together, the advantage of the TS potential is relatively minor, and clearly is offset by the disadvantage that it needs a factor of about 100 more computing time than the BKS potential (which clearly is the second-best choice with respect to accurate prediction of physical properties): This conclusion is corroborated by an investigation of other crystal structures (β-cristobalite, stishovite) [49, 63]. Hence in the following we shall focus on the use of the BKS potential only, since under the circumstances described above it is the most sensible compromise to use the BKS potential.

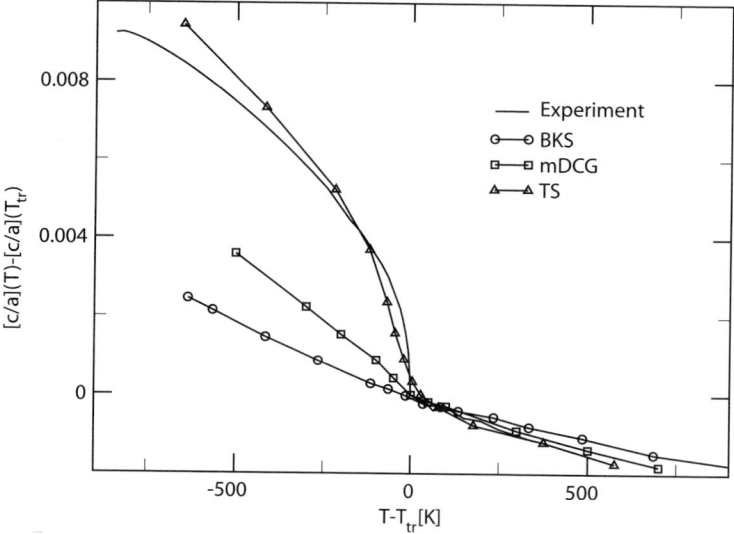

Fig. 1.9. c/a ratio of quartz, relative to its value at the $\alpha \to \beta$ phase transition, plotted vs. the temperature difference from the transition temperature. *Full curves without symbols* show the experiment [35], *curves with symbols* refer to the three potentials studied (cf. Fig. 1.8). From Herzbach *et al.* [49]

1.4 Simulations of Molten and Glassy Silicon Dioxide

Already in the previous section we have discussed the problem of proper equilibration, which is particularly stringent for liquid silica since its extremely large viscosity means that the structural relaxation time is very large as well, and exceeds the time scale accessible to molecular dynamics simulations at the temperatures of interest in experiments by many orders of magnitude. However, in Fig. 1.2 it was already shown that equilibrating the melt at $T = 2750\,\mathrm{K}$ and subsequent cooling at constant volume does produce useful results [43] which compare favorably with experiment [46]. Therefore, it is of interest to discuss such results in greater detail.

A quantity of great interest for a glass is its specific heat. However, there we have to be careful: if we would compute the specific heat directly from the MD simulation in the NVT ensemble, i.e. using either the temperature derivative of the internal energy or the corresponding fluctuation relation

$$C_V = \left(\frac{\partial (\langle \mathcal{H} \rangle / N)}{\partial T} \right)_V = \frac{\langle \mathcal{H}^2 \rangle - \langle \mathcal{H} \rangle^2}{N k_B T^2}, \tag{1.50}$$

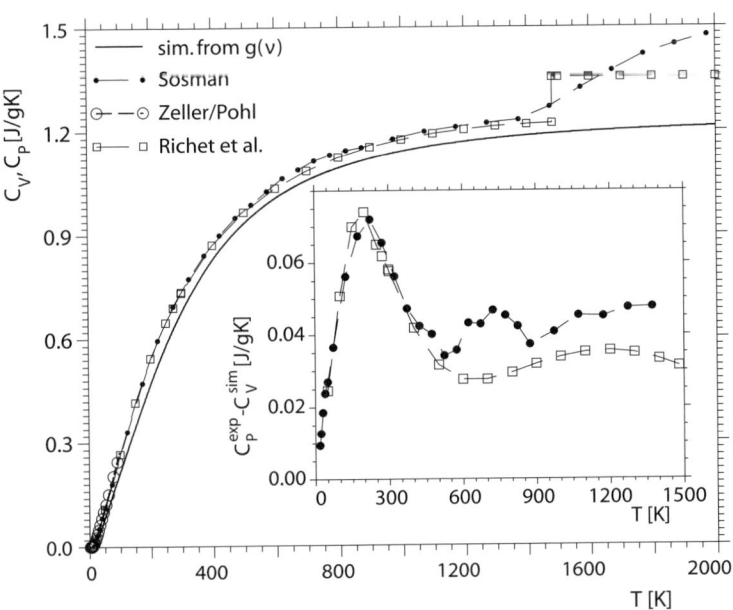

Fig. 1.10. Temperature dependence of the specific heat of SiO_2 as predicted by the harmonic approximation, using the density of states $g(\omega)$ estimated from MD simulations of the BKS model and Eq. 1.51 (*solid line*). The symbols are experimental data for the specific heat at constant pressure from Sosman [64], Zeller and Pohl [65] and from Richet *et al.* [66]. The inset shows the difference between the data of Richet et al. and of Sosman and the simulation data. From Horbach *et al.* [44]

then we would get the Dulong–Petit law at low temperatures, $C_V = 3k_B$, due to the classical statistical mechanics implied by the MD method. This is again a violation of the third law of statistical mechanics, that requires $C_V \to 0$ as $T \to 0$. Since a PIMD simulation of glassy silica would be a major effort, a "cheap" alternative method was applied that is valid when the system at low temperature is strictly harmonic: then one can obtain the vibrational density of states $g(\omega)$ from the classical MD simulation (this was done at $T = 30\,\mathrm{K}$ [44]) and use it in the quantum-mechanical formula for the specific heat

$$C_V = \frac{\hbar^2}{k_B T^2} \int_0^\infty \frac{\omega^2 \exp(\hbar\omega/k_B T)}{[\exp(\hbar\omega/k_B T) - 1]^2} g(\omega) d\omega . \tag{1.51}$$

In this way, quantum mechanics is introduced into the problem "by hand" via the use of the occupation number of states according to the Bose–Einstein statistics. Of course, this method is rather approximate for glasses, where anharmonic effects are present (such as the famous "two-level systems" which are held responsible for the linear specific heat $C_V \propto T$ of glasses at very low temperature) but nevertheless this is a reasonable approximation at temperatures

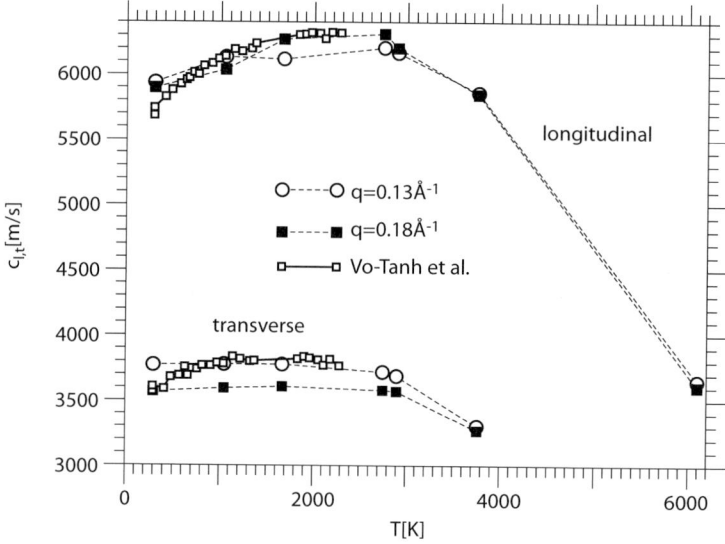

Fig. 1.11. Longitudinal (c_ℓ) and transverse (c_t) sound velocities of SiO$_2$ plotted versus temperature. These quantities were determined from the frequencies at which the maxima of the corresponding longitudinal and transverse current correlation functions at $q = 0.13\,\text{Å}^{-1}$ (*open circles*) and $q = 0.18\,\text{Å}^{-1}$ (*filled squares*) occur. Also included are the corresponding experimental data of Polian *et al.* [68] which are multiplied with the factor $(2.2/2.37)^{1/2}$ since the simulation was done at a density $\rho_{\mathrm{sim}} = 2.37\,\mathrm{g/cm}^3$ while the experiment was done for $\rho_{\mathrm{exp}} = 2.2\,\mathrm{g/cm}^3$ (from Horbach *et al.* [67]

of practical interest, e.g. $T = 300\,\text{K}$. Fig. 1.10 compares the temperature dependence of the specific heat of SiO_2 obtained in this way to corresponding experimental data [64–66]. As expected, the specific heat starts to reach the saturation value of the Dulong–Petit law when T exceeds θ ($\approx 500\,\text{K}$). Surprisingly, the harmonic approximation stays close to the experimental data up to the glass transition temperature ($T_{\text{g}} \approx 1450\,\text{K}$), where the experimental data show a step-like increase when the glass melts. Of course, the harmonic approximation cannot describe the melting of the glass (or of any crystal). In fact, the discrepancies between the simulation and the data are largest for $100\,\text{K} < T < 300\,\text{K}$, and not near T_{g}: this fact suggests that the discrepancies between simulation and experiment are mainly due to an inaccurate estimation of the density of states $g(\omega)$ and not due to anharmonic effects. This conclusion is corroborated by a direct comparison of $g(\omega)$ resulting from the BKS potential with corresponding results derived from CPMD and the TS

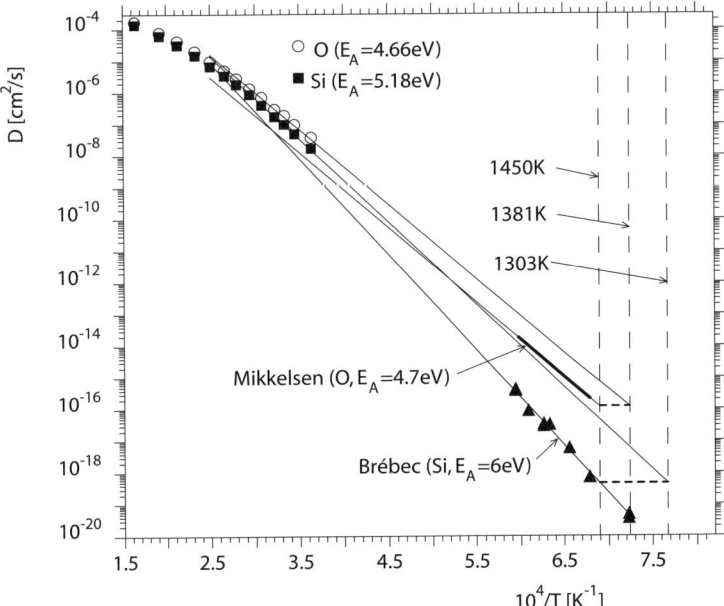

Fig. 1.12. Self diffusion constant D of silicon (Si) and oxygen (O) atoms in molten SiO_2 plotted vs. inverse temperature. The symbols in the upper left part show the MD results, while in the lower right part corresponding experimental data [69,70] are shown. The thin straight lines illustrate Arrhenius relations ($D \propto \exp[-E_A/(k_B T)]$) with activation energies E_A as indicated in the figure. The vertical broken lines show the experimental glass transition temperature, $T_{\text{g}} = 1450\,\text{K}$, as well as estimates for T_{g} obtained from an extrapolation of the simulation results to low T, estimating T_{g} from the experimental value of the O diffusion constant ($D_O(T = T_{\text{sim}}) = 10^{-16}\,\text{cm}^2\text{s}^{-1} \Rightarrow T_{\text{g}}^{\text{sim}} = 1381\,\text{K}$) or the Si diffusion constant, respectively ($D_{\text{Si}}(T = T_{\text{g}}^{\text{sim}}) = 5 \times 10^{-19}\,\text{cm}^2\text{s}^{-1} \Rightarrow T_{\text{g}}^{\text{sim}} = 1303\,\text{K}$). From Horbach and Kob [43]

potential [62]. Also the contributions of the two level systems are only important on the scale of temperatures of the order of $1\,\mathrm{K}$, and hence invisible in Fig. 1.10.

The temperature dependence of the longitudinal (c_ℓ) and transverse (c_t) sound velocity is shown in Fig. 1.11 [67]. These results were obtained from an analysis of the corresponding time-dependent current-current correlation functions, which were Fourier transformed into frequency space. Undamped sound propagation would show up in these correlation functions via δ-functions $\delta(\omega - \omega_q)$ with $\omega = cq$ where $c = c_\ell$ or c_t, respectively [67]. Of course, in a liquid for $q \to 0$ no static shear can be maintained, and hence the transverse mode for $q \to 0$ gets overdamped and ultimately can no longer be identified. Thus in a liquid for $q \to 0$ only c_ℓ is very well-defined. However, for the values of q used in Fig. 1.11 both longitudinal and transverse correlators show

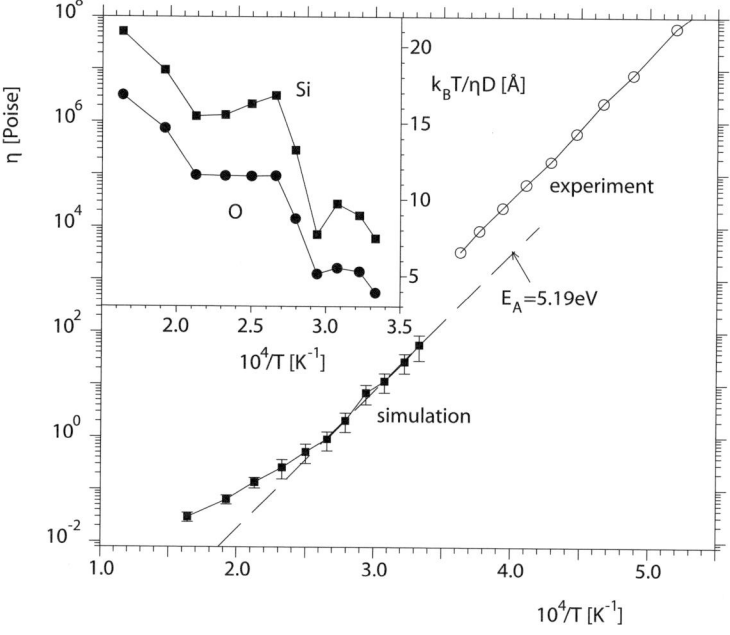

Fig. 1.13. MD results (*filled squares*) for the viscosity of the BKS model for molten SiO_2 plotted versus inverse temperature. The dashed straight line indicates an Arrhenius fit with an activation energy $E_A = 5.19\,\mathrm{eV}$. The experimental data [74] (*open circles*) are compatible with this value but would suggest a slightly larger preexponential factor. Note that for SiO_2 the analysis [43] of other correlation functions at very high temperature suggests a critical temperature $T_c = 3330\,\mathrm{K}$ of mode coupling theory [71], therefore $\eta(T > T_c)$ deviates from the Arrhenius law. The inset shows that the Stokes–Einstein relation $k_B T/(\eta D) = \mathrm{constant}$, where D are the Si or O self diffusion constants, is not valid in the regime of temperatures studied. From Horbach and Kob [43]

broad peaks, and those peak positions are shown in Fig. 1.11, together with corresponding experimental data [68].

While sound waves at nonzero q probe the dynamics of small displacements of the atoms at short times, the self diffusion constants probes the mean square displacements of the atoms over very large time scales (Fig. 1.12) [43]. The Arrhenius plot shown there using scales such that the experimental data [69, 70] can be included on the same plot. From this graph it is quite evident that there is still a broad gap between the temperatures where the simulations can be performed and the temperatures where experiments are available. In fact, the experimental data stop at $T = 2750\,K$, a temperature far above the melting temperature of crystalline SiO_2, so it is not yet possible at all to simulate undercooled SiO_2 melts! Nevertheless it is encouraging, that an extrapolation of the simulations in order to bridge the gap between simulations and experiments yields diffusion constants which are only one or two orders of magnitude off from their experimental counterparts. Furthermore, one can see that at very high temperatures the plot of $\log D$ vs. $1/T$ is clearly curved, indicating that at these high temperatures the simple Arrhenius relation is no longer valid. In fact, a more detailed analysis shows [43] that this regime of temperatures for SiO_2 is described by the mode coupling theory [71], with a critical temperature of $T_c \approx 3330\,K$ [43]. This finding is a very important result, since it shows that SiO_2, the prototype of a "strong" glass-forming liquid, is not fundamentally different from the so-called "fragile" liquids, which are characterized by a non-Arrhenius behavior of their structural relaxation times (see [72, 73] for a more detailed discussion of the implications of these simulations on the theory of the glass transition). The behavior of the shear viscosity, Fig. 1.13, which was extracted from the simulations via the Green–Kubo formula, Eq. (1.41), corroborates these conclusions, and the activation energy that can be extracted from the simulation results agrees well with the corresponding experiment [74]. But we see it as a real strength of the simulations that they can explore regimes of very high temperatures, that are not accessed by the experiments which in this way are complemented by the simulations.

1.5 Mixtures of Silicon Dioxide with Sodium Oxide and Aluminium Oxide

In the previous section, we have shown that MD simulations have given valuable insight into the structure and dynamics of molten and glassy silica. Hence, it is tempting to extend these studies to mixtures of SiO_2 with Na_2O and Al_2O_3, which are basic ingredients of most glasses that are used for various applications. However, here we shall not go into any detail about the studies that have recently begun [75–78], but summarize only very briefly the most salient features.

Again a crucial step is to choose a valid potential. An extension of the BKS potential to include Na and Al was proposed [79], with effective charges $q_{Na} = 1.0$, $q_{Al} = 1.9$. However, this choice does not ensure charge neutrality of the model for general compositions {in practice, one wants to study systems such as $(Na_2O)x(SiO)$, with $x = 2, 3, 5$, for instance, which shall be abbreviated as NSx in the following [75, 76, 78]}. Therefore the extension of the BKS potential [37] due to Kramer *et al.* [79] was slightly modified, introducing distant dependent charges $q_\alpha(r)$ for Na and Al that ensure charge neutrality. E.g., for the case of Na the choice [75]

$$q_{Na}(r) = \begin{cases} \tilde{q}_{Na}\left\{1 + \ln\left[C_{Na}(r_{Na} - r)^2 + 1\right]\right\}, & r < r_{Na} \\ \tilde{q}_{Na} & , \quad r \geq r_{Na} \end{cases} \qquad (1.52)$$

with $\tilde{q}_{Na} = 0.6$, $r_{Na} = 4.9\,\text{Å}$, $C_{Na} = 0.926\,\text{Å}$ was made. Due to Eq. (1.52), $q_{Na}(r)$ increases smoothly from $\tilde{q}_{Na} = 0.6$ for $r \geq r_{Na}$ (this choice ensures charge neutrality, remember $q_O = -1.2$) to $q_{Na} = 1.0$ which is reached for $r = 1.7\,\text{Å}$. A similar modification was also used for $q_{Al}(r)$ [77].

The "sample preparation" was done similarly as in the case of pure SiO_2, i.e. the system was equilibrated in the fluid phase at a given density {taken from experiment, e.g. $\rho = 2.60\,\text{g/cm}^3$ for $(Al_2O_3)2(SiO_2)$ [AS2] for the lowest temperature where full equilibration of the melt still turned out possible on a 10 ns time-scale. It was found that this temperature was $T = 2300\,\text{K}$ for AS2 [77] and $T = 1900\,\text{K}$ for NSx [75]. Cooling then the models down from this state at constant density to the temperatures of interest (e.g. $T = 300\,\text{K}$), again a comparison with suitable experimental scattering data for the (total) structure factor $S(q)$ {as well as other experimental data, as far as available} was used. E.g., in the case of NS2 the simulations (done at $\rho = 2.37\,\text{g/cm}^3$) compare favorably well with the neutron data for $S(q)$ of Misawa *et al.* [80], with a slight exception: the simulations show a small shoulder at $q = 0.9\,\text{Å}^{-1}$, that was not revealed by the experiment. If one would have just the simulation results for $S(q)$ and nothing else, one easily could attribute this "prepeak" at $q = 0.9\,\text{Å}^{-1}$ to statistical noise, and hence doubt that it is a real effect. However, here crucially the advantage comes into play that the simulation can go beyond experiment, i.e. record quantities that are not accessible in the experiment, such as the (six) partial structure factors $S_{\alpha\beta}(q)$ [75–78]. One finds [75, 76, 78] that $S_{NaNa}(q)$ has a very pronounced maximum at $q \approx 0.9\,\text{Å}^{-1}$, but both $S_{NaO}(q)$ and $S_{NaSi}(q)$ have minima at $q \approx 0.9\,\text{Å}^{-1}$, therefore in the total structure factor $S(q)$ shows only a weak shoulder there. It turns out that this structural feature can be linked [76] to the existence of a network of percolating Na-rich channels in the structure, and these channels provide the explanation why it is possible that the Na diffusion constant is several orders of magnitude larger than the diffusion constants of oxygen or silicon, respectively [75, 81]. The existence of such channels has been hypothesized in the literature for a long time, but clear evidence has been lacking and thus the issue was somewhat controversial, until the MD results [76] did settle the issue. Gratifyingly, the finding of the "prepeak" at

$q = 0.9\,\text{Å}^{-1}$ in the simulations also prompted experimental colleagues to carry out extensive new neutron scattering experiments [81,82], which subsequently proved the existence of such a prepeak in the actual scattering data from the real material as well. This is a nice example illustrating the fruitful interplay between simulations and experiments.

While the introduction of Na_2O into SiO_2 melts has the effect to break up the continuous random network of Si–O covalent bonds, creating O–ions near Na ions with a single covalent bond to a Si ion (i.e. "dangling bonds" are created), the introduction of Al_2O_3 has a very different effect: all Al_2O_3 molecules are incorporated into the network of covalent bonds. However, due to the different stoichiometry Al_2O_3 and SiO_2 do not really fit together, if the rules formulated above for an ideal SiO_2 network are obeyed (each Si has four O neighbors, each O is neighbor of two Si ions). As a consequence, there is the need to form "triclusters", i.e. oxygens which are bond to an Al ion must have three rather than two covalent bonds. Again the simulations provide striking evidence for this effect [77], and show at the same time that the mobility of Al ions is not enhanced in comparison with Si or O ions, unlike the Na case, since now all ions (Al, Si, O) are part of the same network. Nevertheless, also in this case a peak at small q is found in the partial structure factors $S_{\text{AlAl}}(q)$ and $S_{\text{SiSi}}(q)$, indicating a tendency towards microphase separation between Al-rich and Si-rich nanoscopic regions in the melt (actually undercooled melts at much lower temperatures than are accessible by simulations show a true miscibility gap in the phase diagram). Thus, also in the case of SiO_2–Al_2O_3 mixtures the MD simulations have yielded a wealth of interesting and nontrivial information, allowing again for a better understanding of this system.

1.6 Conclusions

Although the title of these lecture notes is "Computer Simulations of Undercooled Fluids and Glasses", an important conclusion is that for archetypical glass formers such as molten SiO_2 and its mixture with Na_2O, Al_2O_3 etc. it is not yet possible to simulate such undercooled fluids in metastable equilibrium (i.e., at temperatures below their melting temperature T_m). In view of the fact, that the structural relaxation time of a glassforming fluid increases by about 15 orders of magnitude, from 10^{-12} s to 10^3 s, when the glass transition temperature T_g is approached, this caveat should not be too surprising, since MD is most powerful for studying dynamics in the range from 10^{-12} s to 10^{-7} s, and at T_m the structural relaxation time for many glass forming fluids is already much larger than 10^{-7} s. Developing methods that would allow MD work to cover many more decades in time without excessive brute-force use of computer resources in fact is one of the "grand challenge" problems for computer simulation in the future.

Nevertheless it has been shown that the study of glass forming fluids in the temperature range where they can be equilibrated allows very useful insights, helps to interpret experimental data which otherwise would be ambiguous, and allows to make useful extrapolations that bridge the time-scale gap almost quantitatively. Also predictions of properties of the glass can be made, despite the disparity in the cooling rates between simulation and experiment (however, effects of these huge cooling rates need to be carefully considered!).

Of course, when qualitative insight into phenomena is not the primary goal of the study, but rather a quantitative account of properties of real materials, the accuracy of the available effective potentials is a crucial problem. Such potentials are best tested not for fluids but for the corresponding crystals, as demonstrated here by a comparative study of several potentials for α-quartz and β-quartz. Then it is important to take care of quantum effects at low temperatures, (e.g. possible by PIMD) and of finite size effects near structural phase transitions (possible by finite size scaling). Addressing all these issues, we have shown that an effective potential that is both simple and perfectly accurate clearly does not exist, but sometimes (like in the case of SiO_2) simple potentials such as the BKS potential are an affordable compromise.

Acknowledgement. Very valuable contributions to the research reviewed here are due to W. Kob, K. Vollmayr, and A. Winkler: it is a pleasure to thank them. We also like to thank SCHOTT Glas and the Bundesministerium für Bildung und Forschung (BMBF) for financial support (BMBF grant No 03N6015). We thank the John von Neumann Institue (NIC) at Jülich and the Höchstleistungsrechenzentrum Stuttgart (HLRS) for generous allocations of computer time on the CRAY-T3E.

References

1. M.P. Allen, D.J. Tildesley: *Computer Simulation of Liquids* (Clarendon Press, Oxford, 1987)
2. K. Binder, G. Ciccotti (Eds.): *Monte Carlo and Molecular Dynamics of Condensed Matter Systems* (Societá) Italiana di Fisica, Bologna, 1996)
3. D. Frenkel, B. Smit: *Understanding Molecular Simulation: From Algorithms to Applications* (Academic Press, San Diego, 1996)
4. D.P. Landau, K. Binder: *A Guide to Monte Carlo Simulations in Statistical Physics* (Cambridge Univ. Press, Cambridge, 2000)
5. J. Grotendorst, D. Marx, M. Muramatsu (Eds.): *Quantum Theory of Complex Many-Body Systems: From Theory to Algorithms – Lecture Notes* (NIC, Jülich, 2002)
6. N. Attig, K. Binder, H. Grubmüller, K. Kremer (Eds.): *Computational Soft Matter: From Synthetic Polymers to Proteins – Lecture Notes* (NIC, Jülich, 2004)
7. M. Karttunen, I. Vattulainen, A. Lukkarinen (Eds.) *Novel Methods of Soft Matter Simulations. Lecture Notes in Physics, Vol 640* (Springer, Berlin, 2004)
8. K. Kremer, G.S. Grest: J. Chem. Phys. **92**, 5057 (1990)

9. K. Binder (ed.): *Monte Carlo and Molecular Dynamics Simulations in Polymer Science* (Oxford University Press, New York, 1995)
10. K. Binder, D.W. Heermann: *Monte Carlo Simulations in Statistical Physics. An Introduction (4th Ed.)* (Springer, Berlin, 2002)
11. A. Sariban, K. Binder: J. Chem. Phys. **86**, 5859 (1987)
12. S.K. Das, J. Horbach, K. Binder: J. Chem. Phys. **119**, 1547 (2003)
13. K. Binder, J.L. Lebowitz, M.K. Phani, M.H. Kalos: Acta Metall. **29**, 1655 (1981)
14. R. Car, M. Parrinello: Phys. Rev. Lett. **55**, 2471 (1985)
15. W. Kohn: In: *Monte Carlo and Molecular Dynamics of Condensed Matter Systems* ed. by K. Binder, G. Cicotti (Societá Italiana di Fisica, Bologna, 996) p. 561; R. Car: *ibid.* p. 601
16. B.J. Berne, D. Thirumalai: Annu. Rev. Phys. Chem. **37**, 401 (1986)
17. D.M. Ceperley: Rev. Mod. Phys. **67**, 279 (1995)
18. D.M. Ceperley: In: *Monte Carlo and Molecular Dynamics of Condensed Matter Systems* ed. by K. Binder, G. Ciccotti (Societa Italiana di Fisica, Bologna, 1996) p. 443
19. P. Nielaba: In: *Annual Reviews of Computational Science, Vol. 5* ed. by D. Stauffer (World Scientific, Singapore, 1997) p. 137
20. D. Marx, M.H. Müser: J. Phys.: Condens. Matter **11**, R117 (1999)
21. M.E. Tuckerman, B.J. Berne, G.J. Martyna, M.L. Klein: J. Chem. Phys. **99**, 2796 (1993)
22. M.E. Tuckerman, A. Hughes: *Classical and Quantum Dynamics in Condensed Phase Simulation* (World Scientific, Singapore, 1998)
23. M.H. Müser: J. Chem. Phys. **114**, 6364 (2001)
24. P. Schöffel, M. H. Müser: Phys. Rev. B **63**, 224180 (2001)
25. D. Marx: Science **303**, 634 (2004); D. Marx, M.E. Tuckerman, G.J. Martyna: Comput. Phys. Commun. **118**, 166 (1999); D. Marx, M. Parrinello: Z. Phys. B **95**, 143 (1994)
26. L. Verlet: Phys. Rev. **159**, 98 (1967)
27. H.C. Andersen: J. Comput. Phys. **52**, 24 (1983)
28. M. Tuckerman, G.J. Martyna, B.J. Berne: J. Chem. Phys. **97**, 1990 (1992)
29. J.L. Lebowitz, J.K. Percus, L. Verlet: Phys. Rev. **153**, 250 (1967)
30. S. Nose: J. Chem. Phys. **81**, 511 (1984)
31. W.G. Hoover: Phys. Rev. A **31**, 1695 (1985)
32. C. Bennemann, W. Paul, K. Binder, B. Dünweg: Phys. Rev. **57**, 843 (1998)
33. H.C. Andersen: J. Chem. Phys. **72**, 2384 (1980)
34. L.P. Kadanoff, P.C. Martin: Ann. Phys. (N.Y.) **24**, 419 (1963)
35. J.-P. Hansen, I.R. McDonald: *Theory of Simple Liquids* (Academic, San Diego, 1986)
36. J.P. Boon, S. Yip: *Molecular Hydrodynamics* (McGraw Hill, New York, 1980)
37. B. van Beest, G. Kramer, R. van Santen: Phys. Rev. Lett. **64**, 1955 (1990)
38. S. Tsuneyuki, M. Tsukada, H. Aoki, F. Matsui: Phys. Rev. Lett. **61**, 869 (1988)
39. K. Vollmayr, W. Kob, K. Binder: Phys. Rev. B**54**, 15808 (1996)
40. C.A. Angell, J.H.R. Clark, L.V. Woodcock: Adv. Chem. Phys. **48**, 397 (1981)
41. J. Horbach, W. Kob, K. Binder: Phil. Mag. B **77**, 297 (1998)
42. J. Horbach, W. Kob, K. Binder: J. Non-Cryst. Solids **235-237**, 320 (1998)
43. J. Horbach, W. Kob: Phys. Rev. B **60**, 3169 (1999)
44. J. Horbach, W. Kob, K. Binder: J. Phys. Chem. B **103**, 4104 (1999)
45. J. Horbach, W. Kob, K. Binder, A.C. Angell: Phys. Rev. E **54**, R5897 (1996)

46. D.L. Price, S.M. Carpenter: J. Non-Cryst. Solids **92**, 153 (1987)
47. P.J. Heaney, C.T. Prewitt, G.V. Gibbs (eds.): *Silica – Physical Behavior, Geochemistry and Materials Applications* (Mineral Soc. Am., Washington D.C., 1994)
48. M.H. Müser, K. Binder: Phys. Chem. Minerals **28**, 746 (2001)
49. D. Herzbach, K. Binder, M.H. Müser: J. Chem. Phys. **123**, 124711 (2005)
50. M. Parrinello, A. Rahman: Phys. Rev. Lett. **45**, 1196 (1980)
51. K. Binder: In: *Phase Transformations in Materials* ed. by G. Kostorz. Wiley-VCH, Weinheim (2001), p. 239
52. P.C. Hohenberg, B.I. Halperin: Rev. Mod. Phys. **49**, 435 (1977)
53. K. Binder: In: *Computational Methods in Field Theory* ed. by H. Gausterer and C.B. Lang (Springer, Berlin, 1992), p. 59
54. K. Binder: Rep. Progr. Phys. **60**, 487 (1997)
55. M.A. Carpenter, E.K.H. Salje, A. Graeme-Barber, B. Wruck, M.T. Dove, K.S. Knight: Am. Mineral **83**, 2 (1998)
56. K. Binder: Z. Phys. B **43**, 119 (1981)
57. M.E. Fisher: Rev. Mod. Phys. **46**, 587 (1974)
58. H. Grimm, B. Dorner: J. Phys. Chem. Solids **36**, 407 (1975)
59. J.S. Tse, D.D. Klug: J. Chem. Phys. **95**, 9176 (1991)
60. S. Tsuneyuki, H. Aoki, M. Tsukuda: Phys. Rev. Lett. **64**, 776 (1990)
61. E. Demiralp, T. Cagin, W.A. Goddard III: Phys. Rev. Lett. **82**, 1708 (1999)
62. P. Tangney, S. Scandolo: J. Chem. Phys. **117**, 8898 (2002)
63. D. Herzbach: *Dissertation* (Johannes Gutenberg Universität Mainz, 2004)
64. R.B. Sosman: *The Properties of Silica*, (Chemical Catalog Co., New York, 1927)
65. R.C. Zeller, R.O. Pohl: Phys. Rev. B**4**, 2029 (1971)
66. P. Richet, Y. Bottinga, D. Denielou, J.P. Petitet, C. Tegui: Geochem. Cosmochim. Acta **46**, 2639 (1982)
67. J. Horbach, W. Kob, K. Binder: Eur. Phys. J. B **19**, 539 (2001)
68. A. Polian, D. Vo-Tanh, P. Richter: Europhys. Lett. **57**, 375 (2002)
69. J.C. Mikkelsen: Appl Phys. Lett. **45**, 1187 (1984)
70. G. Brébec, P. Seguin, C. Sella, J. Bevenot, J.C. Martin: Acta Metall. **28**, 327 (1970)
71. W. Götze, J. Sjögren: Rep. Prog. Phys. **55**, 214 (1992)
72. W. Kob: J. Phys.: Cond. Matter **11**, R85 (1999)
73. K. Binder, W. Kob: *Statistical mechanics of disordered materials* (World Scientific, Singapore, 2005)
74. G. Urbain: Geochim. Cosmochim. Acta **46**, 1061 (1982)
75. J. Horbach, W. Kob, K. Binder: Chem. Geol. **174**, 87 (2001)
76. J. Horbach, W. Kob, K. Binder: Phys. Rev. Lett. **88**, 125502 (2002)
77. A. Winkler, J. Horbach, W. Kob, K. Binder: J. Chem. Phys. **120**, 384 (2003)
78. K. Binder, J. Horbach, W. Kob, A. Winkler: In: *Complex Inorganic Solids III* ed. by P.E.A. Turchi et al., (Springer, Berlin, 2005), p. 35
79. G.J. Kramer, A.J.M. de Man, R.A. van Santen: J. Am. Chem. Soc. **64**, 6435 (1991)
80. M. Misawa, D.L. Price, K. Suzuki: J. Non-Cryst. Solids **37**, 85 (1980)
81. A. Meyer, H. Schober, D.B. Dingwell: Europhys. Lett. **59**, 708 (2002)
82. A. Meyer, J. Horbach, W. Kob, F. Kargl, H. Schober: Phys. Rev. Lett. **93**, 027801 (2004)

Simulation of Inorganic Nanotubes

Andrey N. Enyashin, Sibylle Gemming, and Gotthard Seifert

Summary. Motivated by the high application potential of carbon nanotubes, the search for other quasi one-dimensional nanostructures has been pursued both by theoretical and experimental approaches. The investigations soon concentrated on layered inorganic materials, which may be exfoliated and rolled up to tubular and scroll-type forms. The present chapter reviews the basic design principles, which govern the search for novel inorganic nanostructures on the basis of energy- and strain-related stability criteria. These principles are then applied to the prediction and characterisation of the properties of non-carbon, elemental and binary nanotubes derived from layered boride, nitride, and sulfide bulk phases. Finally, the present chapter introduces examples, where one-dimensional nanostructures such as tubes and scrolls have successfully been constructed from non-layered materials, especially from oxides. Examples for the experimental verification of the predicted structures are given throughout the discussion and impressively underline the predictive power of today's materials modelling.

2.1 Introduction

Extensive investigations of the nanostructured matter began in 1991 after the discovery of tubular forms based on the graphite layers with diameters in order of a few nanometers [1]. The new one-dimensional (1D) allotropic modification of carbon was denoted as nanotube.

Nanotubes have attracted much attention among scientists from both experiment and theory. The information about the construction, the synthesis, the properties of the carbon nanotubes and of the related $B_xC_yN_z$ nanotubes was published in numerous publications and books [2–4,9–12]. At present this new modification of carbon is not only in the focus of various fundamental investigations but also finds a widespread application, e.g. as light receptors and emitters [5,6], in parallelised scanning probe techniques [7], or, more recently in nanoelectronic transistors.

Simultaneously with the discovery of carbon nanotubes a question appeared – can nanotubular structures be produced only from carbon? The synthesis of non-carbon nanotubes based on molybdenum and tungsten disulphides in 1992 demonstrated, that hollow 1D nanostructures are not exclusively phenomena of carbon but can also be synthesised from other compounds [13]. During 15 years the nanotubulenes of dozens of inorganic compounds were discovered. Like carbon nanostructures some of them were found in rocks from earth and space [14, 15]. In contrast to the carbon containing *"organic"* $B_xC_yN_z$ nanotubes, these nanostructured hollow particles were termed *inorganic* nanotubes.

Although the first inorganic nanotubes were observed one year after the carbon nanotubes, the majority of investigations on inorganic nanotubulenes is dedicated to new synthesis routes. The physical and chemical properties of inorganic nanotubes are less well studied; some information was published in the reviews of Refs. [16–24]. A common finding is that the properties of these nanoobjects differ from the ones of the stable bulk modification and may be interesting not only for scientists but also have a technological perspective like the carbon nanotubes. Therefore at present the theoretical modeling of the stability and properties of nanotubes plays an important role and is motivating the necessity of a further detailed experimental study of these phenomena.

The present overview gives the basic results of the theoretical modeling of the structural and electronic properties of the inorganic nanotubes, as developed in the last years.

2.2 Design of Inorganic Nanotubes

As carbon nanotubes, most of the inorganic nanotubes known at present were synthesized from layered – quasi two-dimensional – bulk compounds. The characteristic peculiarity of these compounds is the clearly expressed anisotropy of the strong (covalent) and weak (Van-der-Waals') bondings within and between the layers, respectively. Experimentally obtained nanotubes usually have complex structures: they can be composed of various numbers of coaxial cylindrical layers and the bonds within the layers exhibit different orientations relative to the tube axis. Nanotubes can be open or closed at the ends. They can have not only cylindrical, but also a scroll-like morphology. Nevertheless, all the tubular forms of matter may be structurally characterised by using the classification developed for the single-walled nanotubes, which are produced by rolling up of monolayer.

The structure of the crystalline monolayer can be characterized using the terms of a two-dimensional (2D) lattice spanned by the translation vectors (a_1 and a_2) and the angle between them (φ). Five types of 2D Bravais lattices exist (Fig. 2.1): oblique ($a_1 \neq a_2$, $\varphi \neq 90°$), square ($a_1 = a_2$, $\varphi = 90°$), hexagonal ($a_1 = a_2$, $\varphi = 120°$), primitive rectangular ($a_1 \neq a_2$, $\varphi = 90°$), and centered rectangular ($a_1 \neq a_2$, $\varphi = 90°$). Cylindrical surfaces may be constructed by rolling up these lattices.

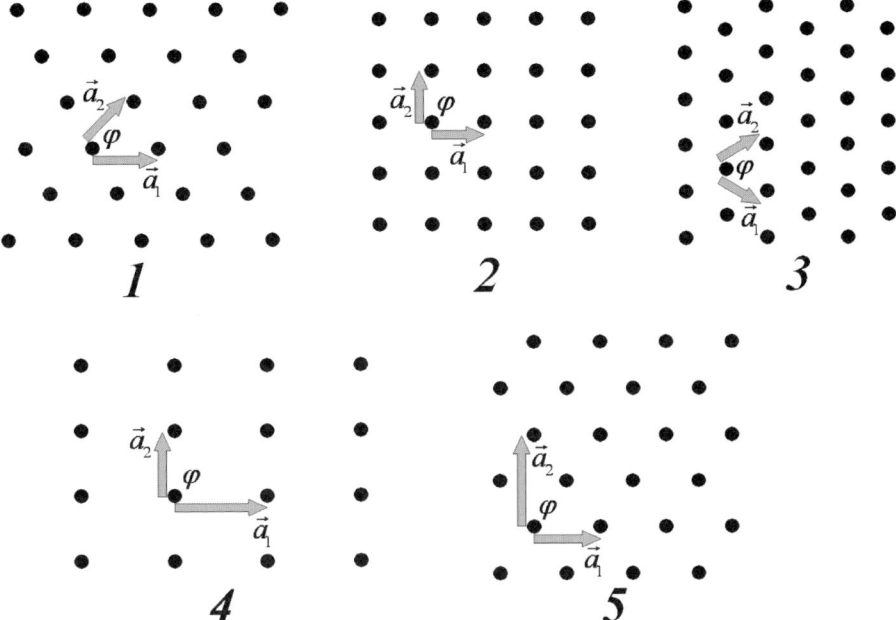

Fig. 2.1. Possible types of the 2D Bravais lattices

The walls of carbon nanotubes are composed of carbon hexagons with the same ordering as in the hexagonal graphenic layer. The basic principles of the geometry specification of the "ideal" carbon nanotubes used at present are motivated and developed in more detail in Ref. [2].

A lot of layered inorganic compounds, which were discovered to form nanotubes, also exhibit layers with a hexagonal atom arrangement (Fig. 2.2). Therefore the classification of the carbon nanotubes may often be used for the classification of the inorganic nanotube structure. Every cylindrical nanotube is characterized by a radius R and by a type of the helicoidal atomic arrangement (chirality) determined by a chiral angle θ. Using the basis vectors of the 2D hexagonal lattice $a_1 = a_2$ and the chiral vector $c = na_1 + ma_2$ it is possible to describe the basic geometry parameters of a tubulene – R and θ, which is produced by rolling of a ribbon sliced from a monolayer (Fig. 2.3):

$$\theta = \arctan \frac{\sqrt{3}m}{m + 2n}, \tag{2.1}$$

$$R = \frac{|c|}{2\pi} = \frac{|a|}{2\pi} \sqrt{3(n^2 + m^2 + mn)}. \tag{2.2}$$

Since R and θ are associated uniquely with a_1 and a_2 using the integer indexes n and m, these indices may be used for the structural classification of the inorganic nanotubes in the same way as for carbon nanotubes – (n, m).

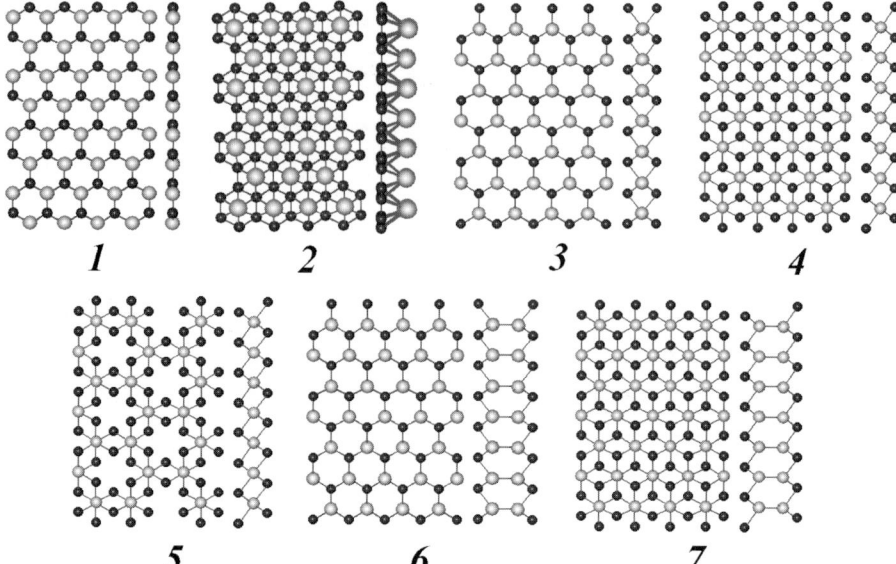

Fig. 2.2. Structures of the hexagonal monolayers of some (quasi)layered compounds: *1* – BN; *2* – AlB$_2$, MgB$_2$; *3* – MoS$_2$, WS$_2$, NbSe$_2$; *4* – TiS$_2$, ZrS$_2$, anatase TiO$_2$, NiCl$_2$; *5* – FeCl$_3$, MoO$_3$; *6* – GaS, GaSe; *7* – metastable InS

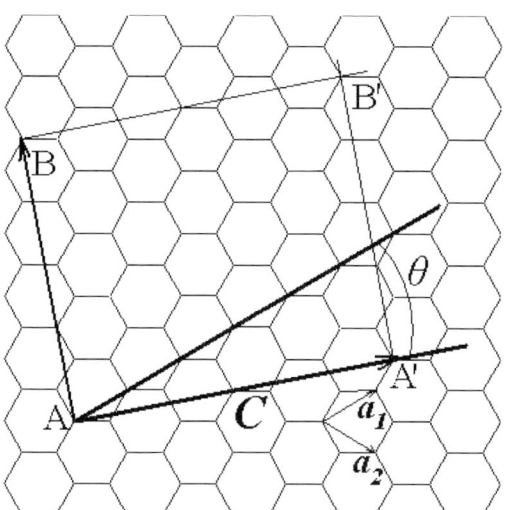

Fig. 2.3. The way of construction of a single-layered nanotube based on a hexagonal layer

Then, depending on the values of n and m all such nanotubes can be subdivided into two groups: chiral tubes with $0° < \theta < 30°$ and the non-chiral ones, the so-called *"zigzag"* and *"armchair"* tubes with $\theta = 0°$ and $30°$, respectively.

Single-walled tubular structures based on nonhexagonal layers may analogously be described by the primitive vectors of the respective lattice (Fig. 2.4). In those cases, R and θ will be associated with n and m by other correlations.

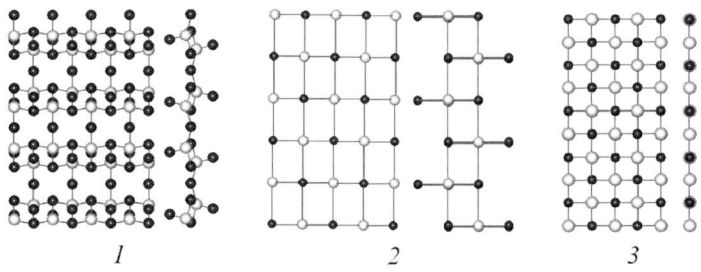

Fig. 2.4. Structures of the nonhexagonal monolayers of some (quasi)layered compounds: *1* – rectangular V_2O_5; *2* – centered rectangular lepidocrocite TiO_2; *3* – square MgO

The symmetry of the nanotubes determines many of their physical properties connected particularly with electronic and phonon states. A detailed review for tubes based on hexgonal layers, such as MS_2, (M = Mo, W), XN (X = B, Ga), C, BC_3, or BC_2N, is given in Refs. [25, 26].

It is important to note, that single-walled carbon nanotubes are based on a planar, almost ideally two-dimensional graphitic layer, which is just one carbon atom thick. In contrast, in most of the layered inorganic compounds a single layer has a more complicated atom arrangement and is several atoms thick. Therefore, inorganic single-walled tubes are composed of several concentric and interconnected cylinders of atoms (Fig. 2.2 and 2.4). This fact gives *a priori* a basis to suppose, that inorganic tubulenes are less stable than carbon ones with the same radii, since a rolling of a thicker layer to a cylinder is energetically less preferable.

The model monolayers of the inorganic compounds described so far represent only a small part of the possible structural variety. Some other models will be viewed in the next parts devoted to the modeling of inorganic nanotubes.

The inorganic nanotubes as carbon tubes are often closed at the ends by "caps". At present structure models of these "caps" are offered for the tubes based on the hexagonal layers of the graphitic-like BN or AlB_2 types or on the layers of the metal dichalcogenides and dichlorides with prismatic or octahedral coordination of metallic atoms (Fig. 2.5). Similarly to the carbon

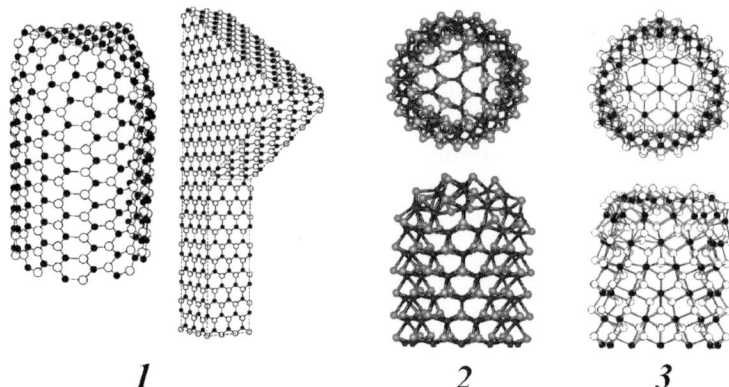

Fig. 2.5. The "capped" single-walled inorganic nanotubes with structure of:
1 – BN [27]; *2* – MoS$_2$ [28]; *3* – ZrS$_2$ [29]

nanostructures they are fragments of the corresponding hollow molecules, the inorganic fullerenes, which may connect to the end of a nanotube using one of the various possible defects of the tube wall, for example the nonhexagonal cycles.

2.3 General Criteria for the Stability of Inorganic Nanotubes

The fundamental energetic parameter of the system stability is the total energy (E_{tot}). For a comparative analysis of the nanotube stability usually one employs the strain energy (E_{str}), which is determined as the difference of the total energies of the nanotube and the corresponding infinite monolayer.

At present two approaches have been developed to assess the nanotube stability – the continuum and the atomistic ones.

First of them are the phenomenological models based on the principles of the classical theory of elasticity [30]. In the framework of this approach the atomic structure of single-walled nanotubes is not taken in account, and the dependence of E_{str} on the radius R is given by

$$E_{str} = \frac{\pi Y L h^3}{12R},\tag{2.3}$$

where h – thickness of the monolayer, L – length of the nanotube, Y – elastic module of the monolayer. The number of atoms in the nanotube wall is defined as $N = 2\pi R L \varrho_a$, where ϱ_a is the number of atoms on the surface unit of monolayer. Then the strain energy per atom is equal to

$$\frac{E_{str}}{N} = \frac{Y h^3}{24\rho_a R^2}.\tag{2.4}$$

A thermodynamic model was created for two-phase multi-layered films and nanotubes in Ref. [31]. It was shown, that the dominance of one or another nanoform is determined by the balance between the surface and interfacial energies, the energies of interactions between the interfaces and interfaces and surfaces, and the bending energies.

In atomistic approaches the quantitative estimates of the energetic parameters are realized in the framework of the molecular mechanics or quantum-mechanics methods. As the continuum model they give qualitatively the same dependence of the strain energy on the radius $E_{str} \sim 1/R^2$. Thus, among nanotubes with the same radius the tubes formed from the thicker monolayers are the less stable ones. For example, the strain energies of the carbon nanotubes formed from the monoatomic layer are compared to the strain energies of the TiS_2 and GaSe tubes formed with three- and tetraatomic monolayers, respectively [32, 33] (Fig. 2.6).

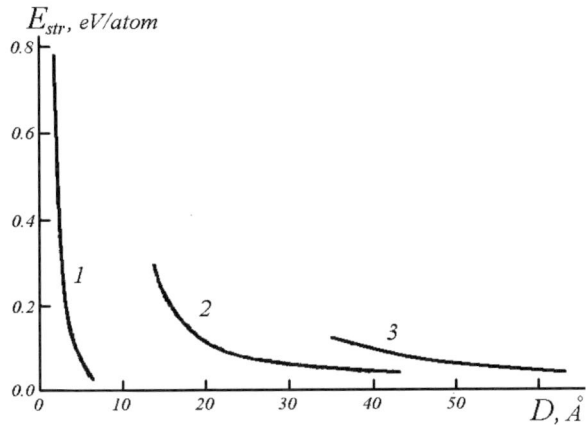

Fig. 2.6. The comparison of the strain energies of the carbon nanotubes (*1*) to those of the inorganic nanotubes TiS_2 (*2*) and GaS (*3*) [32, 33]

The results of quantum simulations allow to derive energetic models of stability of the various nanostructures – nanotubes, nanostrips, nanorolls and fullerenes [34]: E_{str} may be represented as the sum of the atom-based energy contributions from atoms on the surface with the non-zero curvature and from atoms with dangling bonds. For the multi-layered nanostructures the contribution of the Van-der-Waals' interaction between surfaces of the neighboring layers should additionally be taken in account. This simple speculation is very efficient for the comparative study of the stability, because the possibilities of the quantum methods are limited by the size of the studied system. The elaboration of such models allows to judge the stability of the real obtained nanosystems. In particular, the energetic model of the cylindric MoS_2 nanotubes and the plain MoS_2 nanostrips offered in Ref. [34] shows that nanotubulenes are more stable than strips if $R > 6\,\mathrm{nm}$, which is in the agreement with the experimental data. In accordance with experimental data, multi-layered nanosystems are more stable than the mono-layered ones owing to attractive interlayer Van-der-Waals' interactions.

In the discussion of the stability of inorganic nanotubulenes it should be noted, that the majority consists of compounds with considerably covalent bonds. This finding may be explained using a semiquantitative model for the calculation of the coulombic energy of a nanotube as function of the ionicity [35]. A direct estimate of the Coulomb interaction energy by using the Madelung constant for every nanotube with its unique chirality is a very complicated problem. However, it is easy to estimate the electrostatic energy of a nanotube as a function of R if we take a nanotube as an ideal cylindrical capacitor. For example, for stoichiometric MX_2 tubulenes with octahedral or trigonal prismatic coordination of atoms M the Coulomb contribution to the strain energy of the tubulenes $E_{\mathrm{str}}^{\mathrm{ion}}$ may be estimated as

$$
\frac{E_{\mathrm{str}}^{\mathrm{ion}}}{N} = \frac{Z^2 e^2 V_{\mathrm{mol}}}{6\varepsilon_0 x^4 w} \left(R \ln \left(\frac{R+l}{R-l} - 2l \right) \right) \approx \frac{Z^2 e^2 V_{\mathrm{mol}}}{18\sqrt{2}\varepsilon_0 xw} \frac{1}{R^2} , \tag{2.5}
$$

where e – electron charge, e_0 – dielectric constant, Z – charge on atom M, V_{mol} and w – molar volume and the interlayer distance (thickness of the monolayer) in the bulk material MX_2, x – average bond length M–X and l is equal $x/\sqrt{2}$. Equation 2.5 demonstrates that the high ionicity of a compound does not promote the stabilization of the nanotubulenes based on it. It is interesting, that the ionic part $E_{\mathrm{str}}^{\mathrm{ion}}$ of the strain energy E_{str} is proportional $1/R^2$, too.

2.4 Theoretical Prediction of the Properties of Non-Carbon Nanotubes

Most of the theoretical predictions of the properties of the nanotubes are carried out using band theory models considering a nanotube as an ideally one-dimensional (1D) system. This approximation is justified as long as the diameters of nanotubes are much smaller than their lengths. The unit cell of binary, ternary or even quaternary inorganic nanotubes usually contains more atoms than a carbon nanotube with the same chirality, thus the computational effort is increased. The solution of this problem was found in the application of modern electronic structure methods like the density functional theory (DFT) or semiempirical methods together with periodic boundary conditions along the tube. Due to the advances in the numerical implementation of DFT and to the increase of the compute power also larger cluster models can nowadays be treated, especially for a study of the nanotube ends. The structural characteristics, the kinetics of formation and the thermal properties are studied also using molecular dynamics methods. The details of some approaches and the calculation schemes may be found in the original articles discussed below.

It should be noticed, that despite available data on the synthesis of inorganic nanotubes based on non-layered bulk compounds the subjects of the theoretical investigations are limited usually to compounds with a stable or a metastable layered phase. Some results from the modeling of stability and properties of nanotubulenes are presented below ordered by the chemical elements and their chemical compounds.

2.4.1 Nanotubes of the IVA Group Elements

The IVa element most related to carbon is silicon, but the possibility of a synthesis of pure silicon nanotubulenes, which are topologically identical to graphenic tubes, is disputable for electronic structure reasons. For a Si atom the sp^2-configuration is not stable, thus, in contrast to carbon, the most stable structure of silicon is the diamond structure.

First calculations of the electronic states of the hypothetical silicon nanotubes based on hexagonal layers were carried out using an ab-initio method [36, 37] for the (6,6), (10,0), and (8,2) tubes. The behaviour of the electronic properties of such tubulenes is similar to the ones of carbon nanotubes. Particularly, the *zigzag* nanotube is a semiconductor with narrow band gap, and the *armchair* nanotube exhibits a metallic energy spectrum. The total energy difference between bulk Si and the graphitic-like layer or the nanotube shows that bulk silicon is more stable by $0.79\,\mathrm{eV/atom}$ and $0.83\,\mathrm{eV/atom}$, respectively. Thus, the strain energy of the silicon nanotubulenes is nearly the same as for the carbon tubes, but the formation of a graphite-like layered silicon structure is highly unfavourable compared with carbon. These conclusions and the dependence of the electronic properties on the chirality are confirmed by other *ab initio* and semiempirical calculations of nanotubular Si clusters; furthermore, the calculations agree that silicon tubulenes would be based on a buckled cyclohexane-type layer rather than on a planar honeycomb network [38, 39]. According to [39] the *armchair* tube configuration is the most stable one due to the more effective overlap of the p-orbitals and the delocalization π-bonds.

The thermal and mechanical properties were investigated using the empirical Tersoff potential. As expected, the affinity of the Si atoms to sp^3-hybridization leads to smaller temperatures of a desintegration in comparison to those of the carbon nanotubes [37]. The behaviour of Si nanotubes under tension and torsion is similar to the behaviour of carbon nanotubes, but the mechanical characteristics are essentially different [40, 41]. For example, the Young moduli of the silicon tubulenes are lower than the ones of carbon tubulenes by a factor of $10 - 20$.

Like silicon also germanium crystallises in the diamond structure. Germanium nanotubes based on graphitic-like layers were not investigated, but probably such structures should be even less stable than the ones of silicon. Buckled hexagonal nets of Si or Ge atoms could, however, be obtained by a chemical stabilisation of the sp^3-state, e.g. by saturation of the free valences by hydrogen atoms or other elements, as shown in refs. [43,44] for silane SiH and germane GeH nanotubulenes (Fig. 2.7). The stability of those structures was investigated with the DF-TB method. As for other types of tubes the strain energies vary proportionally $1/R^2$. But in contrast to the carbon and silicon nanotubes the band gap does not depend on the chirality and changes monotonously with the radius. All these nanotubes are semiconductors. The band gap values of the SiH and GeH tubes with the largest radii studied are equal 2.50 eV and 1.33 eV respectively. The typical value of the Young modulus of the SiH tubes is about 0.08 TPa, i.e. by a factor of 15 lower than for the carbon tubulenes [45].

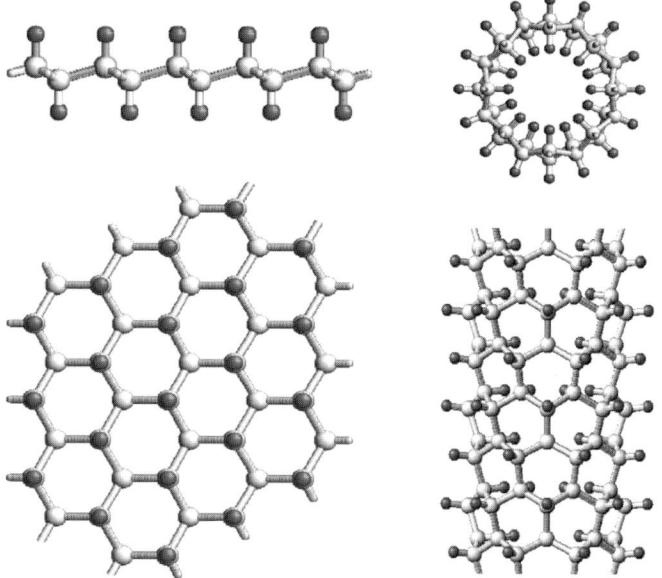

Fig. 2.7. A monolayer of silane SiH and (8,0) SiH nanotube [43]

Bundles of (6,6) nanotubulenes based on the quasilayered compound calcium disilicide CaSi$_2$ were considered for example by density-functional band structure calculations [46]. In all cases the structure of the Si skeleton is very puckered. Despite the strongly ionic character of bonding between the Si and Ca shells these nanotubes have a metallic-like core and, accordingly a metallic density of states. A similar situation was obtained for the related CaAlSi and SrGaSi nanotubes, where one half of the Si atoms is substituted by Al or Ga atoms [47]. The results of the semiempirical EHT method calculations show the existence of a strong covalent bonding within the AlSi (GaSi) skeleton and a weak interaction with the Ca(Sr) atoms. All these nanotubes have the metallic conductivity.

2.4.2 Nanotubes of the VA Group Elements

The most stable allotropes of phosphorus, arsenic, antimony, and bismuth are the layered modifications. These elements are isomorphic. Each atom connects pyramidally to the three neighboring atoms in the same layer and has the three close contacts to atoms in the neighboring layer. On the whole, the structure of these layers is similar to the graphite, but all hexagonal cycles are deformed and have the "armchair" conformation.

The structural, electronic and mechanical properties of hypothetical chiral nanotubulenes based on black phosphorus were predicted by DFTB calculations [30,48]. The strain energy of the phosphoric nanotubes changes proportionally $1/R^2$, but is higher than for the carbon tubes with the same radii. It is caused by the increasing repulsion of the lone pairs when rolling the P layer to a cylinder. The typical value of the Young modulus is equal 0.3 TPa. All phosphorus nanotubes are semiconductors with band gaps of $0.9 - 1.2$ eV.

In good qualitative and quantitative agreement with the above mentioned data are the results of first-principles DFT calculations [49] carried out for chiral and nonchiral tubulenes and nanostrips of phosphorus. It was confirmed that all P nanotubes are semiconductors and that the band gap is determined by the chirality, but in all cases its value aspires to the value of the P monolayer 1.8 eV. Furthermore, all tubes of phosphorus are more stable than the initial strips if the radius is larger than 0.55 nm.

An *ab-initio* molecular dynamics method was used for the study of the stability and electronic properties of bismuth nanotubes [50]. The values of the strain energies of these tubulenes are very close to the values of the carbon tubes and changes according to the classical dependence on R. All Bi nanotubes are semiconductors and for large diameters the band gap converges to the value of 0.63 eV obtained for the monolayer at Γ-point. In accordance with these results the synthesis of Bi nanotubes was reported [42].

Taking into account the similarity of the properties of phosphorus and arsenic, antimony, and bismuth the existence of As and Sb nanotubulenes may not be excluded [48].

2.4.3 Nanotubes of Boron and Borides

Boron forms numerous three-dimensional structures, thus numerous cluster calculations on boron structres have been carried out. One of the first studies of a tubular boron structure is an *ab-initio* RHF-SCF-based investigation of the isomers of B_{32} (spherical, cone-shaped, quasiplanar and tubular) [51]. The surface of the spherical and cone-shaped clusters is formed by hexagonal B_7 and pentagonal B_6 pyramids, and the most stable structures have the lowest curvature. Even more stable are the quasiplanar and tubular clusters, constructed only from hexagonal pyramids, in which the elimination of dangling bonds determines the stability. Also smaller clusters like B_{24} were investigated using molecular dynamics within the framework of DFT [52].

Different theoretical methods were applied to calculate the properties of boron nanotubulenes both for cluster models and for periodically repeated cells [53,54]. The electronic structure of all considered tubulenes has metallic-like character and does not differ from the one of the boron monolayer. The strain energies of the boron nanotubes are comparable with the strain energy of other known – carbon and boron nitride – nanotubes, i.e. a synthesis of boron nanotubulenes should be quite possible.

New results [54] of *ab-initio* calculations on boron nanotubes show that these systems may have non-cylindrical forms. In particular for bundles of the highest tubulenes polygonal conformer were obtained. The edges are based on the icosahedral B_{12} fragments, typical for the crystalline allotropes of boron. These structures may be interpreted as intermediates between ideal tubular phases and the known bulk phases.

The structures including network of boron atoms may be stabilized by substitution with other elements. For example by exchanging the tops of the hexagonal pyramids B_7 in the puckered boron layer by metal atoms one can get the structure of the layered borides of AlB_2 type (Fig. 2.8). In this manner the electron-accepting boron layer is stabilised by the partial transfer of electrons from the metal atoms. As for the silicide tubes the chemical bonds are covalent within the boron layer and ionic between the boron and metal atoms. For the first time the stability and morphology of hypothetical diboride nanosystems were considered using a molecular mechanics method [55]. Using a first-principles DFT method AlB_2 nanotubulenes [56] were investigated, which can be obtained by rolling of the layers of the bulk AlB_2 and the metallic atoms may be placed both inside and outside of a boron cylinder. It was found, that all studied nanotubes should have metallic type of the electroconductivity as the boron nanotubes.

The structure, the nature of chemical bonding and electronic properties of the single- and multi-layered tubulenes based on MgB_2, AlB_2, ScB_2 and TiB_2 borides were analyzed using band calculations within framework of the semiempirical Extended Hueckel (EHT) method [57–59]. All considered tubulenes have metallic-like electronic spectra independent of the composition and the number of layers. The most stable tubes are these, where the metal atoms

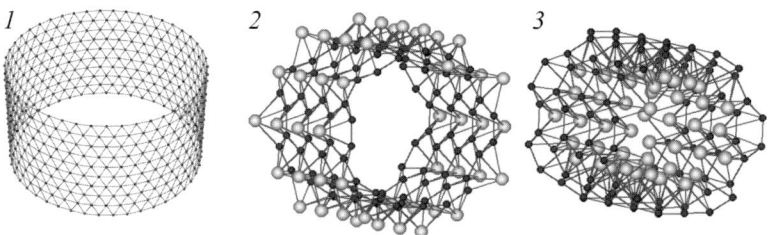

Fig. 2.8. The boron-containing nanotubes: *1* – (15,15) B [43]; *2,3* – (6,6) MgB_2 with metal atoms outside and inside of the boron cylinder [57].

are located inside of the boron skeleton. The results of the first-principles calculations of the stoichiometric $(6,0)TiB_2$ and composite $(6,0)TiB_2@(12,0)B$ nanotubes were published in Ref. [60]. It was found, that the strain energy of the $(6,0)TiB_2$ tube is equal about $0.09\,eV/atom$, which is comparable to the analogical value of the carbon nanotubes, thus a synthesis of this tube may be achievable.

2.4.4 Nanotubes of Boron Nitride and its Analogues

The layered modification of boron nitride BN was the first non-carbon compound predicted theoretically as suitable for the construction of inorganic nanotubes. Using various quantum methods several important characteristics of these nanotubes were predicted, which have been confirmed experimentally. At present BN nanostructures are studied in nearly as much detail as the carbon nanostructures.

Particularly it was found, that – in contrast to carbon nanotubes – the electronic properties of the boron nitride tubulenes do not depend on the chirality [61]. All BN tubes have a wide band gap of about $5 - 6\,eV$, decreasing with the decrease of the tube radius. The *zigzag* tubulenes are semiconductors with a direct band gap $(\Gamma - \Gamma)$, and other nanotubes have an indirect band gap $(\Delta - \Gamma)$.

The dependence of the electronic properties on the structures and the sizes is less pronounced than for the carbon nanotubes. This circumstance is very important for the synthesis of nanomaterials with required parameters.

The BN nanotubes are the first inorganic nanotubes for which an influence of various defects was considered theoretically. As in the case of the carbon nanotubes the defects can be caused by doping of the walls, by formation of endo- and exohedral structures, or by structural defects of the walls.

E.g. the first simulation of heterostructures based on BN nanotubes with intercalated Al or K atoms was reported in Ref. [62]. The energies of intercalation for the $(4,4)BN$ tube are much less than for the similar carbon tube. The $Al@(4,4)BN$ composite has additional $Al s, p$ levels in the band gap. At the same time the electronic spectrum is very close to a superposition of the spectra of the Al monoatomic chain and $(4,4)BN$ nanotube, which indicates the insignificant interactions between them. The charge density distribution for the near-Fermi states demonstrate, that the leading density of the conductance electrons of this composite concentrates inside of the tube. This behaviour differs from the ones of the carbon nanotubes, where the dominating contribution to conductivity occurs at the tube walls. Hence, the BN nanotubes may be more preferable as a winding of the nanocables. A little more complicated variant of the nanocable based on a metallic armchair carbon nanotube as the core and a BN nanotube as the winding was offered in Ref. [63]. As it was expected the electronic structure of this nanocable is the superposition of the electronic spectra of its components, and its conductivity should be caused by the properties of the core.

The electronic properties of the BN nanotube doped with a single impurity atom were studied for example for tubes doped with silicon [64]. Using an *ab-initio* method both variants of substitution Si_B and Si_N were investigated. In both cases an impurity atom protrudes outwards. The formation of the Si_B defect was found energetically more preferable. The Fermi level in this case lies at the defect level of the impurity atom Si. For the BN nanotube doped by Si_N defect the levels of the impurity are localized in the band gap of the tube.

Also the mechanical characteristics of the boron nitride nanotubes are investigated in sufficient detail. The analysis of the elastic properties of the single-walled BN nanotubes in comparison to the characteristics of the carbon and mixed $B_xC_yN_z$ nanotubes shows their lower toughness [65]. E.g. the Young modulus is equal to about $0.8 - 0.9\,TPa$, compared with the corresponding values of $1.2 - 1.3\,TPa$ for carbon tubes and of about $1.0\,TPa$ for BC_2N tubes. Thus, the energy for the rolling of a planar BN nanostrip to a nanotube does not differ much from the rolling energy of a graphite nanostrip.

The structural and thermal properties of BN nanotubes have been investigated using molecular dynamics methods with empirical potentials. In particular, the Tersoff potential was employed to calculate strain energies, which agree semiquantitatively with the data of first-principles calculations [66]. The study of the thermal properties show, that the temperature of the tube destruction depends on the configuration and size. The *armchair* tubulenes are destroyed at higher temperatures than the *zigzag* tubulenes with the same radii. The nanotubes with smaller radii disintegrate at the lower temperatures, because they have higher strain energies. An important role in the process of the thermal degradation of BN nanotubes is played by the burning of the non-hexagonal cycles in their walls similar to the Stone–Wales defects in the carbon nanotubes. Evidently the formation of these defects is a key step for the mechanical destructions of the BN nanotubes, as demonstrated by an *ab-initio* simulation in the framework of the cluster model [67].

The number of the theoretical and experimental investigations of the inorganic BN nanotubes and other BN nanostructures is large. For a more detailed information the reader is referred to refs. [17–19, 23, 68]. No doubt, the boron nitride nanostructures together with the carbon nanoforms will be not only a sufficiently simple and obvious material for a study of the nanosized form of matter, but also will find a wide technological application in the near future.

Besides the nanotubes of boron nitride the existence of the tubulenes based on another nitrides – the isoelectronic analogues – nitrides of aluminium AlN and gallium GaN have been considered. Numerous calculations and experiments show the big difference between the energies of the quasi-2D and 3D modifications of these compounds compared with BN. Nevertheless the rolling energy of the hexagonal AlN monolayer is theoretically lower than for the carbon and inorganic nanotubes based on the monoatomic layers [69, 70]. Molecular dynamics simulations demonstrate, that these AlN nanotubes are stable at room temperature and begin to disintegrate at $600\,K$. Similarly to

the boron nitride nanotubes the AlN nanotubes are wide gap semiconductors
with band gaps of $2.9 - 4.0\,\mathrm{eV}$, increasing with the increase of the radius. As
BN tubes the AlN tubulenes of *zigzag* configuration are semiconductors with
a direct band gap and the *armchair* nanotubes have an indirect band gap.

Density-functional calculations were carried out for the investigation of
GaN nanotubes based on the graphitic-like monolayer of gallium nitride [71,
72]. The same trends in the behaviour of the stability and electronic structure
are obtained as discussed above for AlN tubulenes. Taking into account the
importance of GaN in spintronic and optoelectronic devices, the electronic
structure and magnetic characteristics of a cluster model for the (5,5)GaN
nanotube doped by manganese were calculated [73]. The introduction of this
impurity results in additional spin-polarized levels of Mn in the band gap of
the gallium nitride nanotube.

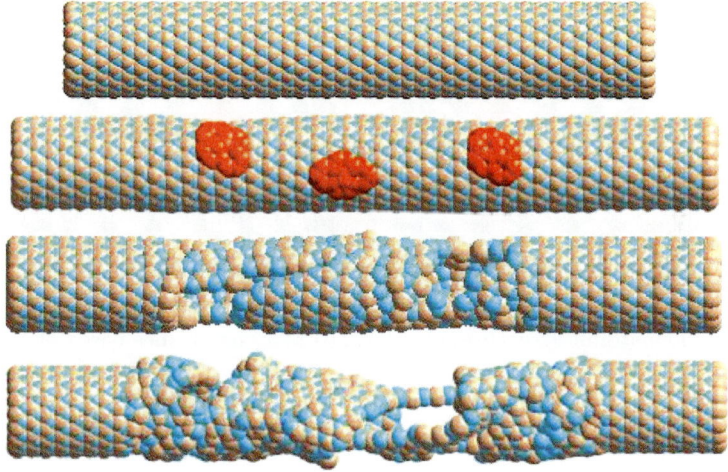

Fig. 2.9. The molecular dynamics simulation of the (9,0)GaN nanotube under
tension [75] © 2004 IOP Publishing Ltd

The empirical Tersoff potential was used for the molecular dynamics modeling of the thermal properties of GaN tubes and their behaviour under tension and fatigue [74, 75]. As for the BN nanotubes a destruction of the GaN
nanotubes begins very fast after the burning of the Stone–Wales defects [75]
(Fig. 2.9).

2.4.5 Nanotubes of Chalcogenides

The nanotubes of metal chalcogenides, namely MoS_2 and WS_2, were first synthesized by Tenne et al. [13], and the first quantum-mechanics-based calculations of these nanotubes were carried out using the DFTB method [28, 76, 77].

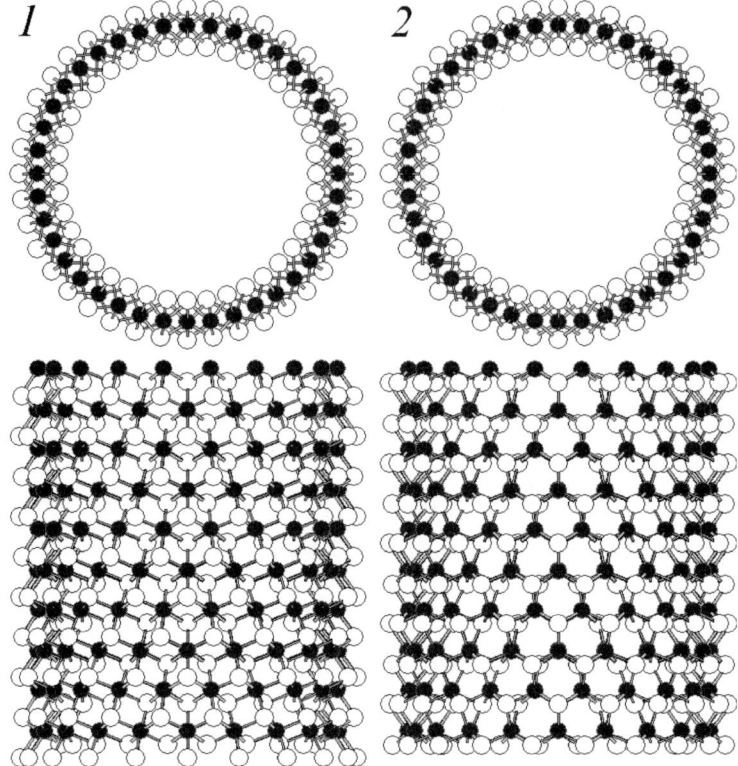

Fig. 2.10. The atomic structures of the (20,0) MX_2 nanotubes based on octahedral (*1*) and prismatic (*2*) units MX_6

Besides, at the same chirality indexes the number of atoms in the unit cell of a chalcogenide nanotube is larger than that in the carbon tube.

The inorganic chalcogenide tubes are produced usually using the layered dichalcogenides MX_2 constructed from MX_6 units with a trigonal prismatic or octahedral coordination of the M atoms (Fig. 2.10). The energetic and electronic characteristics of these tubulenes have got the general features, which are also typical for their bulk phases (Fig. 2.11).

The MoS_2 and WS_2 nanotubes prefer the prismatic coordination of the metal atoms and are the semiconductors with a narrow band gap [28, 76]. Their valence band is composed of three bands represented by contributions of quasi-core S3s-states, hybridized S3p-Mo4d(W5d)-states and Mo4d(W5d)-states. The lower edge of the conduction band is composed by d-states of metals. The *zigzag* tubulenes have a direct transition at the Γ point and the *armchair* tubulenes are semiconductors with an indirect band gap. The value of the band gap increases with the increase of the radius, but this increase is less pronounced than for the planar monolayers.

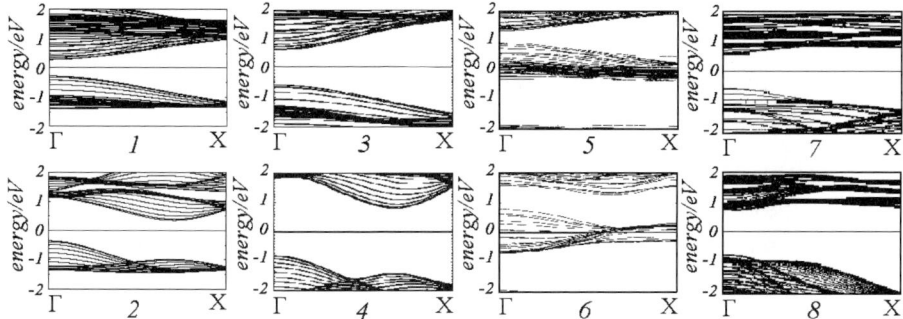

Fig. 2.11. The band structures of the dichalcogenide nanotubes: *1,2* – (22,0) and (14,14) MoS_2 [28]; *3,4* – (22,0) and (18,18) WS_2 [76]; *5,6* – (22,0) and (10,10) NbS_2 [77]; *7,8* – (15,0) and (20,20) TiS_2 [32]

The NbS_2 nanotubes are isostructural to the MoS_2 and WS_2 tubulenes, and they are metallic [77]. The energy bands are represented by the same states but the Fermi level shifts to the band of the Nb$4d$-states with a high density of states. This picture does not differ very much from the ones of the planar NbS_2 monolayer and may serve as an indication for possibly super-conducting properties. The same is true for the electronic properties of nano-tubes based on related $NbSe_2$, which were investigated within framework of the semiempirical EHT method [78].

Because of the large Van-der-Waals' gap all layered dichalcogenides are good matrices for an intercalation. Therefore, their electronic properties may also be modified not only by defects in the atomic arrangement, but also by an introduction of various impurities into the cavities between the layers. Indeed, calculations of the nonstoichiometric (autointercalated) single-walled tubulenes $Nb_{1.25}Se_2$ [79] demonstrate that the introduction of Nb atoms as intercalant of NbS_2 cause changes of the density of states at the Fermi level. Moreover depending on the chirality of nanotubes and on the order of the Nb atoms (outside or inside of a nanotube) the density of states can be both decreased and increased.

Nanotubes based on the semiconducting disulphides with octahedral co-ordination of the metal atoms (TiS_2 and ZrS_2) were considered using the tight-binding methods [29, 32, 80]. The comparative study of the stability of the TiS_2 monolayer and tubulenes with both possible cases of coordination – octahedral and prismatic were carried out [32]. It was demonstrated that in the nanotubular state the octahedral coordination is more stable and is the same as in the bulk material. The electronic structure of these systems is characterised by two clearly expressed valence bands: S$3s$- and Ti$3d$-states. The conduction band is composed mainly of d-levels of the metal atoms. All TiS_2 nanotubes are semiconductors with narrow band gap, and the nature of its (direct or indirect type of transition) is determined by the coordination of the metal atoms and by the chirality of the nanotubes. All nanotubes based on the octahedral TiS_6 units and the *zigzag* nanotubes based on the prismatic TiS_6 unit have a direct gap; the armchair nanotubes based on prismatic TiS_6 units are semiconductors with an indirect transition.

The common picture of the electronic spectrum of ZrS_2 tubulenes is similar to the systems based on TiS_2 [29,80]. The nanotubes are semiconductors and their band gaps decrease with the decreasing radius down to 0.6 eV.

For all above-mentioned chalcogenide nanotubes the mixed ionic-covalent type of bonding is typical. The effective charges are essentially lesser than the formal ones [28]. The strain energy of the tubulenes decrease with increasing radii in the same way as for the other nanotubes considered theoretically. Yet, the strain energy is much bigger than for the carbon nanotubes with the same radii [32]. This fact is explained with the higher energetical expenses for the rolling of a threeatomic layer to a tube than for that of the monoatomic carbon layer. Probably, the thickness of the layer but not the Young modulus plays the prevailing role in the stabilization of these nanotubes (see Eq. 2.4). This conclusion agrees with the calculation of the Young moduli for the single-walled MoS_2 nanotubes [81]. The typical value is equal 0.23 TPa, i.e. six times less than for the carbon tubes.

Nanotubes of the chalcogenides GaS and GaSe based on tetraatomic monolayers are considered theoretically in Refs. [33,82]. It was demonstrated that the strain energies even are much bigger than for the disulphide nanotubes. The Young moduli of the GaS tubulenes are a few times less than for the carbon nanotubes with the same radii. All examined GaS and GaSe nanotubes are semiconductors with band gap values which converge to the values of the planar monolayers with increasenig radii.

2.4.6 Nanotubes of Oxides

The oxides have a bigger ionicity than chalcogenides and a lot of them has not a clearly expressed layered structure. Most of the oxide nanostructures have been synthesised from aqueous solution. The variety of the reaction conditions leads to a wide range of morphologies and properties of the produced structures. Therefore the modeling of the various oxide nanostructures, the comparison of their stability and properties, and the study of the impurity influence are very important for the purposeful control of their functionalization.

The first calculations of the oxide nanotubes were carried out for TiO_2 nanotubes using the semiempirical EHT method [80,83]. This compound has a few allotropes and an anatase monolayer was used for the construction of the model of the titania single-walled nanotubes. The calculated electronic structure reflects the properties of bulk TiO_2. All considered *zigzag* and *armchair* nanotubes are semiconductors with a wide band gap. The values of the band gap are decreasing with the decrease of the radius. The valence band and the low edge of the conductance band are composed of Ti3*d*-O2*p*- and Ti3*d*-states, respectively.

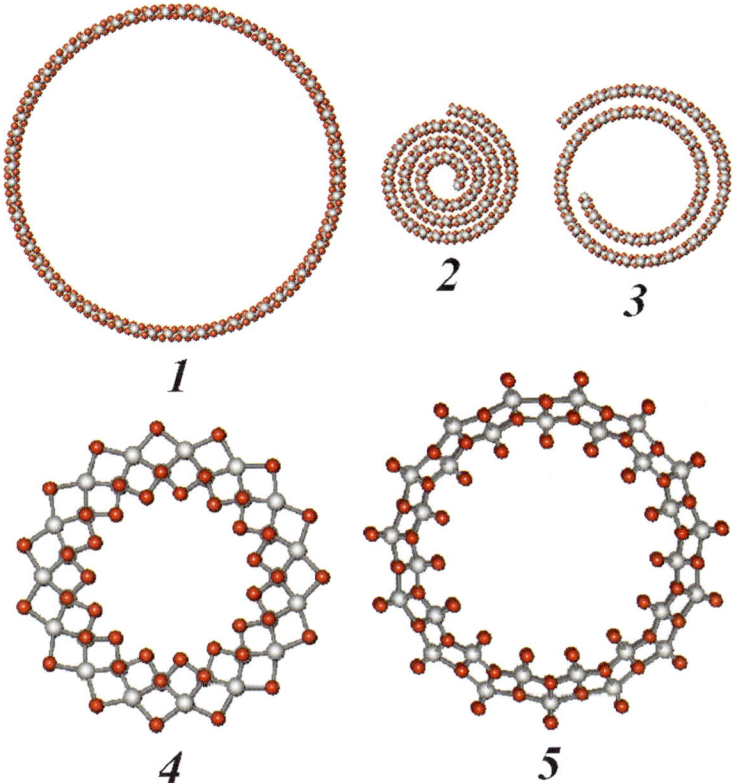

Fig. 2.12. The titania nanostructures: *1* – (60,60) anatase nanotube; *2,3* – (60,60) anatase nanorolls; *4,5* – (15,0) and (0,15) lepidocrocite nanotubes [85]

Using the same method the influence of isoelectronic and heterovalent doping on the electronic structure of titania nanotubes was investigated [84]. Si, Ge, V and Cr atoms were considered as dopants, which substitute one half of the Ti atoms in two ways: first, the "rows" of impurity atoms are aligned along the tube axis and, second, the "rings" are substituted on the tube perimeter. The behaviour of the electronic properties may well be described in the framework of the "rigid" band model. All doped nanotubes with the compositions $Si_{0.5}Ti_{0.5}O_2$ and $Ge_{0.5}Ti_{0.5}O_2$ retain the semiconducting nature and their band gaps change by not more than 0.1 eV. In the case $V_{0.5}Ti_{0.5}O_2$ and $Cr_{0.5}Ti_{0.5}O_2$ nanotubes a partial filling of the low levels of the conduction band takes place and these tubes have metallic-like spectra.

Beside cylindrical anatase tubulenes using the scroll-like nanotubes (nanorolls), the nanostrips of anatase and the nanotubes of another modification – the lepidocrocite TiO_2 [85] (Fig. 2.12). were investigated with the DFTB method. It was established that the cylindrical anatase nanotubes are the

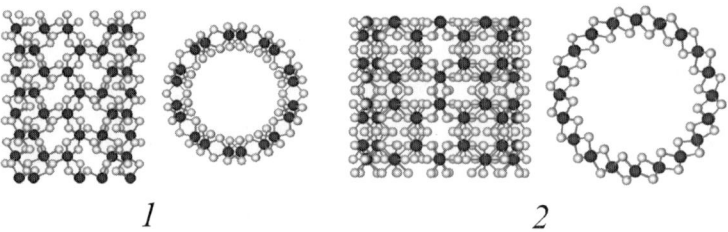

Fig. 2.13. The atomic structures of the (5,5) and (10,0) MoO$_3$ nanotubes [86]

most stable ones among other nanostructures. It is explained by the absence of dangling bonds and the smaller thickness of a layer compared with the TiO$_2$ lepidocrocite. The remarkable conclusion of the electronic structure calculations is the fact, that all considered titanium dioxide nanosystems are semiconductors and their band structure depends weakly on their morphology. Using these quantum-mechanics-based calculations a model of stability of the TiO$_2$ nanorolls was developed.

MoO$_3$ nanotubes were considered in Ref. [86]. Their structure is related to the anatase tubes but has vacancies at the metal atom sites (Fig. 2.13). Semi-empirical calculations demonstrate that they have the same features of the electronic structure as the TiO$_2$ tubulenes.

A very interesting type of inorganic nanotubes are the tubular structures based on vanadium oxide VO$_x$. They differ from "classical" nanotubes. First, their composition depends on the procedure of synthesis and they intercalate organic molecules between layers. Second, cylindrical nanotubes are rare among of the VO$_x$ tubes, but usually they occur as nanorolls. This leads to some problems for the construction of suitable models and a simulation of these systems. At present their electronic properties were examined using simple models based on the divanadium pentoxide V$_2$O$_5$ which has a quasi-layered structure. A V$_2$O$_5$ monolayer has a primitive rectangular lattice and is built by VO$_5$ pyramids. A first attempt of the modeling of the cylindrical V$_2$O$_5$ nanotubes was undertaken in framework of the EHT method [87] (Fig. 2.14). The density of states of the *zigzag* and *armchair* V$_2$O$_5$ nanotubes is similar to the ones of the monolayer and shows that all tubes are semiconductors. The valence band has a mixed V3d-O2p character and is fully occupied. It has three clearly distinguishable bands because of the three nonequivalent types of the oxygen atoms with the different coordination environments. The most intensive band is caused by O2p-states of the vanadyl oxygen. The low edge of the conductance band has contributions of the V3d-states.

The atomic models of the scroll-like V$_2$O$_5$ nanotubes are represented in Fig. 2.14 [88]. Their electronic spectra have the same general characteristics as those for the cylindrical nanotubes. The basic peculiarity is the appearance of additional levels in the band gap, caused by dangling bonds of atoms on the

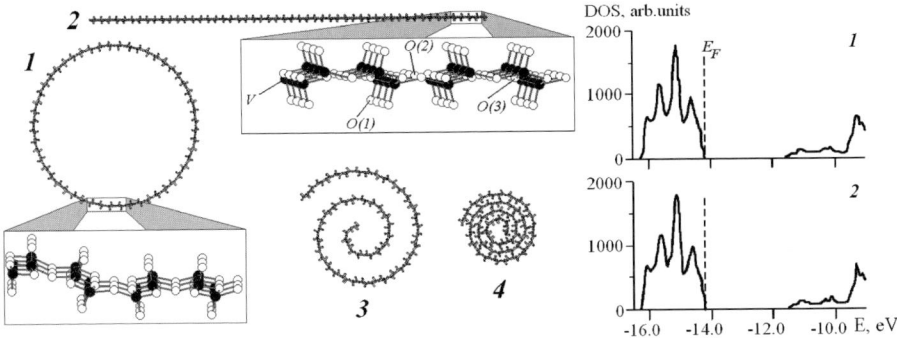

Fig. 2.14. Cross-sectional images of the (26,26) V_2O_5 nanostructures and their electronic densities of states: *1* – nanotube; *2* – nanostrip; *3,4* – nanorolls with two different inter-wall distances [87,88]

internal and external edges of a nanoroll. Depending on the nanoroll chirality and the distance between the neighboring layers the band gap can decrease by up to 0.1 eV. In spite of the presence of dangling bonds the scroll-like V_2O_5 tubes are more stable than the cylindrical ones.

The effects of a doping by Mo atoms on the electronic structure of the vanadium oxide nanotubes were considered in Ref. [89] for the example of the composition $Mo_{0.1}V_{1.9}O_5$ with various variants of the substitution within layer. As for the TiO_2 tubulenes the influence of an electronic doping is well described using the "rigid" band model. The main change of the electronic spectrum is an occupation of the high anti-bonding levels. All tubes obtain metallic spectra. It is supposed, that an increase of the Mo concentration causes the destabilization of these systems and the disintegration of highly-doped nanotubes observed experimentally.

Hypothetical nanotubes of other vanadium oxides – the anatase-like VO_2 were proposed [83]. In contrast to the V_2O_5 nanotubes the general features of their density of states is similar to the spectra of the TiO_2 nanotubes. The main difference consists in the displacement of the Fermi-level to the high-energy band, which results in a metallic-like spectrum of these systems.

For the first time the silicon oxide nanotubes based on a hypothetical siloxene $Si_6H_3(OH)_3$ layer were suggested in Ref. [45]. DF-TB molecular dynamics simulations at 800 K confirm the stability of these tubes. All nanotubes are semiconductors independent of the chirality. The band gap of tubes with large diameters is equal to 1.9 eV. The same authors examined the models of a carbon nanotube covered by a SiO_x monolayer based on the SiO_4 tetrahedra and rolled to a cylinder [90,91]. Three types of interactions are possible depending on x and radius. One of them are covalent Si-C bonds (Fig. 2.15), which realize for nanotubes with large diameter and $x = 5/2$. They lead to a partial rehybridization of the carbon atoms ($sp^2 \rightarrow sp^3$) and to a disturbance of the

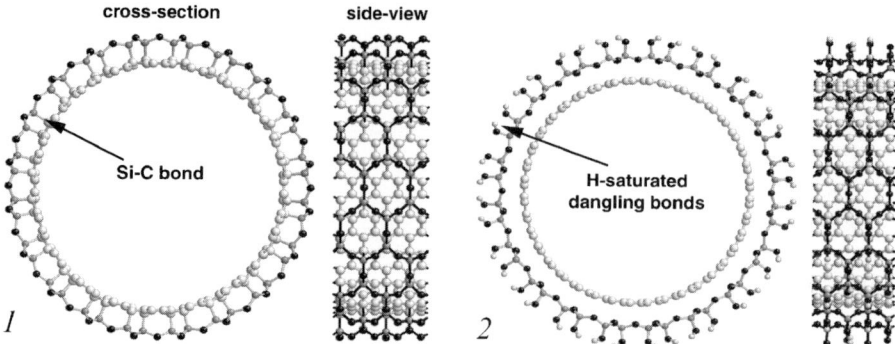

Fig. 2.15. Molecular models of (20,20) carbon nanotubes covered by a cylindrical SiO_x layer ($x = 5/2$) bonded to the carbon tube via Si-C bridges (*1*) and a tubular SiO_x layer ($x = 3/2$) without bonding towards the carbon tube (*2*) [90]

p-electronic system of the caron nanotube. It was established, that the $SiO_{5/2}$ tubes are unstable without a carbon skeleton inside because of the dangling bonds. The second type of interaction is the formation of Si-O-C bonds. This interaction is not advantageous and does not lead to stable structures. The composites of a carbon nanotube and a SiO_x layer are stable, too, without chemical bonds if the dangling bonds of the O atoms are saturated (e.g., by hydrogen atoms).

The cluster models of single-walled MgO nanotubes based on a monolayer of the square lattice were investigated using an *ab initio* method [92]. Similar to the bulk magnesium oxide these nanotubes are insulators. MgO does not have a layered structure and is connected by strong ionic bonds. The more detailed study of these nanotubes may extend the group of the potential precursors for nanotubes synthesis.

As for other nanotubes the general type of the interatomic interactions in the oxide nanotubes is the ionic-covalent bond between metal and non-metal. The stability of all considered oxide nanotubes depends on curvature and the strain energy is proportional to $1/R^2$.

2.5 Conclusion

The material science of the inorganic nanostructures is currently at an initial step of the development. At present the investigation of the properties of inorganic nanotubes is predominantly based on theoretical modeling. However, a precise modeling has to cope with several difficulties, which arise from discrepancies between the idealized, calculated model system and the actually synthesized samples.

The models of nanotubes are represented often as ideal cylinders rolled up from a monolayer, but all experimental data evidence that inorganic nanotubes are multi-layered, and that their shape may deviate from the cylindrical one. In addition, inorganic nanotubular forms from non-layered bulk materials were synthesized, too. Usually inorganic nanotubes grow with numerous structural defects – various vacancies and impurities (from atoms up to large organic molecules). The chemical activity of inorganic nanotubes is currently not well studied by theoretical efforts. Thus, the theoretical investigations concerning the properties of inorganic nanotubes, which are summarized in the present review, represent just a small fragment of the ample research activities on inorganic nanotubes. An overcoming of these problems will be the next achievements in the delivery of the theory of the inorganic nanotubes.

No doubt, the results of the numerous simulations of inorganic nanotubes promote not only a development of our ideas about matter but also will find their incarnation in technological decisions.

References

1. S. Iijima: Nature (London) **354**, 56 (1991)
2. R. Saito, G. Dresselhaus, M.S. Dresselhaus: *Physical Properties of Carbon Nanotubes* (Imperial College Press, London 1998)
3. P.J.F. Harris: *Carbon Nanotubes and Related Structures: New Materials for the Twenty-First Century* (Cambridge University Press, Cambridge 1999)
4. Carbon Nanotubes: Synthesis, Structure, Properties, and Applications and Electronics. In: *Topics in Applied Physics*, vol 80, ed by M.S. Dresselhaus, G. Dresselhaus, Ph. Avouris (Springer, Berlin Heidelberg New York 2001)
5. Y. Miyauchi, S. Chiashi, Y. Murakami, Y. Hayashida, S. Maruyama: Chem. Phys. Lett. **387**, 198 (2004).
6. Y. Wang, K. Kempa, B. Kimball, J.B. Carlson, G. Benham, W.Z. Li, T. Kempa, J. Rybczynski, A. Herczynski, Z.F. Ren: Appl. Phys. Lett. **85**, 2607 (2004).
7. E. Yenilmez, Q. Wang, R.J. Chen, D. Yang, H. Dai: Appl. Phys. Lett. **80**, 2225 (2002)
8. R.V. Seidel, A.P. Graham, J. Kretz, B. Rajasekharan, G.S. Duesberg, M. Liebau, E. Unger, F. Kreupl, W. Hoehnlein: Nano Lett. *ASAP article, DOI: 10.1021/nl048312d.*
9. M. Terrones, W.K. Hsu, H.W. Kroto, D.R.M. Walton: Nanotubes: A Revolution in Materials Science and Electronics. In: *Topics in Current Chemistry*, vol 199, ed by A. Hirsch (Springer, Berlin Heidelberg New York 1999) pp 189–234
10. A.L. Ivanovskii: *Quantum Chemistry in Materials Science: Nanotubular Forms of Matter* (Urals Branch of Russian Academy of Science, Ekaterinburg 1999)
11. C.N.R. Rao, B.C. Satishkumar, A. Govindaraj, M. Nath: Chem. Phys. Chem. **2**, 78 (2001)
12. A.L. Ivanovskii: Russian Chem. Rev. **68**, 103 (1999)
13. R. Tenne, L. Margulis, M. Genut et al.: Nature (London) **360**, 444 (1992)
14. S. Amelinckx, B. Devouard, A. Baronnet: Aca Cryst. A **52**, 850 (1996)
15. T.J. Zega, L.A.J. Garvie, I. Dódony et al.: Serpentine Nanotubes in CM Chondrites. In: *Lunar and Planetary Science Conference*, vol XXXV (League City, Texas) pp 1805
16. R. Tenne: Endeavour **20**, 97 (1996)

17. R. Tenne, A.K. Zettl: Nanotubes from Inorganic Materials. In: *Topics in Applied Physics*, vol 80, ed by M.S. Dresselhaus, G. Dresselhaus, Ph. Avouris (Springer, Berlin Heidelberg New York 2001) pp 81–112
18. V.V. Pokropivny: Powder Metallurgy and Metal Ceramics **41**, 123 (2002)
19. A.L. Ivanovskii: Russian Chem. Rev. **71**, 175 (2002)
20. G.R. Patzke, F. Krumeich, R. Nesper: Angew. Chem. Int. Ed. **41**, 2446 (2002)
21. R. Tenne: Chem. Eur. J. **8**, 5297 (2002)
22. R. Tenne: Angew. Chem. Int. Ed. **42**, 5124 (2003)
23. C.N.R. Rao, M. Nath: Dalton Trans. 1 (2003)
24. M. Remškar: Adv. Mater. **16**, 1497 (2004)
25. I. Milošević, T. Vuković, M. Damnjanović: Eur. Phys. J. B **17**, 707 (2000)
26. M. Damnjanović, T. Vuković, I. Milošević et al.: Acta Cryst. A **57**, 304 (2001)
27. Y. Saito, M. Maida: J. Phys. Chem. A **103**, 1291 (1999)
28. G. Seifert, H. Terrones, M. Terrones et al.: Phys. Rev. Lett. **85**, 146 (2000)
29. V.V. Ivanovskaya, A.N. Enyashin, N.I. Medvedeva et al.: Internet Electr. J. Mol. Des. **2**, 499 (2003)
30. G. Seifert, T. Frauenheim: J. Korean Chem. Soc. **37**, 89 (2000)
31. M.I. Mendelev, D.J. Srolovitz, S.A. Safran et al.: Phys. Rev. B **65**, 075402 (2002)
32. V.V. Ivanovskaya, G. Seifert: Solid State Comm. **130**, 175 (2004)
33. Th. Köhler, Th. Frauenheim, Z. Hajnal et al.: Phys. Rev. B 69, 193403 (2004)
34. G. Seifert, T. Köhler, R. Tenne: J. Phys. Chem. B **106**, 2497 (2002)
35. A.N. Enyashin, Yu.N. Makurin, A.L. Ivanovskii: Doklady Phys. Chem. **399**, 293 (2004)
36. S.F. Fagan, R.J. Baierle, R. Mota et al.: Phys. Rev. B **61**, 9994 (2000)
37. S.F. Fagan, R. Mota, R.J. Baierle et al.: J. Mol. Struct. **539**, 101 (2001)
38. R.Q. Zhang, S.T. Lee, C.-K. Law et al.: Chem. Phys. Lett. **364**, 251 (2002)
39. M. Zhang, Y.H. Kan, Q.J. Zang et al.: Chem. Phys. Lett. **379**, 81 (2003)
40. K.R. Byun, J.W. Kang, H.J. Hwang: J. Korean Phys. Soc. **42**, 635 (2003)
41. J.W. Kang, K.R. Byun, H.J. Hwang: Modelling Simul. Mater. Sci. Eng. **12**, 1 (2004)
42. Y. Li, J. Wang, Y. Wu, X. Jun, D. Yu, P. Yang: J. Am. Chem. Soc. **123**, 9904 (2001)
43. G. Seifert, Th. Köhler, H.M. Urbassek et al.: Phys. Rev. B **63**, 193409 (2001)
44. G. Seifert, Th. Köhler, Z. Hajnal et al.: Solid State Commun. **119**, 653 (2001)
45. G. Seifert, Th. Frauenheim, Th. Köhler et al.: Phys. Stat. Sol. (b) **225**, 393 (2001)
46. S. Gemming, G. Seifert: Phys. Rev. B **68**, 075416 (2003)
47. I.R. Shein, V.V. Ivanovskaya, N.I. Medvedeva et al.: JETP Letters **76**, 189 (2002)
48. G. Seifert, E. Hernández: Chem. Phys. Lett. **318**, 355 (2000)
49. I. Cabria, J.W. Mintmire: Europhys. Lett. **65**, 82 (2004)
50. C. Su, H.-T. Liu, J.-M. Li: Nanotechnology **13**, 746 (2002)
51. I. Boustani, A. Rubio, J.A. Alonso: Chem. Phys. Lett. **311**, 21 (1999)
52. S. Chacko, D.G. Kanhere, I. Boustani: Phys. Rev. B **68**, 035414 (2003)
53. I. Boustani, A. Quandt, E. Hernández et al.: J. Chem. Phys. **110**, 3176 (1999)
54. J. Kunstmann, A. Quandt: cond-mat 0410761 (2004)
55. L.A. Chernozatonskii: JETP Letters 74, **369** (2001)
56. A. Quandt, A.Y. Liu, I. Boustani: Phys. Rev. B **64**, 125422 (2001)

57. V.V. Ivanovskaya, A.N. Enyashin, A.A. Sofronov et al.: J. Mol. Struct. (Theochem) **625**, 9 (2003)
58. V.V. Ivanovskaya, A.N. Enyashin, A.A. Sofronov et al.: Theor. Exp. Chem. **39**, 1 (2003)
59. V.G. Bamburov, V.V. Ivanovskaya, A.N. Enyashin et al.: Doklady Phys. Chem. **388**, 43 (2003)
60. S. Guerini, P. Piquini: Microelectronics J. **34**, 495 (2003)
61. A. Rubio, J.L. Corkill, M.L. Cohen: Phys. Rev. B **49**, 5081 (1994)
62. A. Rubio, Y. Miyamoto, X. Blase et al.: Phys. Rev. B **53**, 4023 (1996)
63. A.N. Enyashin, G. Seifert, A.L. Ivanovskii: JETP Letters **80**, 608 (2004)
64. S. Guerini, T. Kar, P. Piquini: Eur. Phys. J. B **38**, 515 (2004)
65. E. Hernández, C. Goze, P. Bernier et al.: Phys. Rev. Lett. **80**, 4502 (1998)
66. W.H. Moon, H.J. Hwang: Nanotechnology **15**, 431 (2004)
67. T. Dumitrica, H.F. Bettinger, G.E. Scuseria et al.: Phys. Rev. B **68**, 085412 (2003)
68. N.G. Chopra, A. Zettl: Boron-nitride-containing Nanotubes. In: *Fullerenes: Chemistry, Physics, and Technology*, ed by K.M. Kadish, R.S. Ruoff (John Wiley & Sons 2000) pp 767–794
69. M. Zhao, Y. Xia, D. Zhang et al.: Phys. Rev. B **68**, 235415 (2003)
70. M. Zhao, Y. Xia, Z. Tan et al.: Chem. Phys. Lett. **389**, 160 (2004)
71. S.M. Lee, Y.H. Lee, Y.G. Hwang et al.: J. Korean Phys. Soc. **34**, S253 (1999)
72. S.M. Lee, Y.H. Lee, Y.G. Hwang et al.: Phys. Rev. B **60**, 7788 (1999)
73. S. Hao, G. Zhou, J. Wu et al.: Phys. Rev. B **69**, 113403 (2004)
74. J.W. Kang, H.J. Hwang, K.O. Song et al.: J. Korean Phys. Soc. **43**, 372 (2003)
75. Y.-R. Jeng, P.-C. Tsai, T.H. Fang: Nanotechnology **15**, 1737 (2004)
76. G. Seifert, H. Terrones, M. Terrones et al.: Solid State Comm. **114**, 245 (2000)
77. G. Seifert, H. Terrones, M. Terrones et al.: Solid State Comm. **115**, 635 (2000)
78. V.V. Ivanovskaya, A.N. Enyashin, N.I. Medvedeva et al.: Phys. Stat. Sol. (b) **238**, R1 (2003)
79. A.N. Enyashin, V.V. Ivanovskaya, I.R. Shein et al.: Russian J. Struct. Chem. **45**, 579 (2004)
80. V.V. Ivanovskaya, A.N. Enyashin, A.L. Ivanovskii: Russian J. Inorg. Chem. **49**, 244 (2004)
81. I. Kaplan-Ashiri, S.R. Cohen, K. Gartsman et al.: J. Mater. Res. **19**, 454 (2004)
82. M. Côté, M.L. Cohen, D.J. Chadi: Phys. Rev. B 58, R4277 (1998)
83. V.V. Ivanovskaya, A.N. Enyashin, A.L. Ivanovskii: Mendeleev Commun. **13**, 5 (2003)
84. A.N. Enyashin, V.V. Ivanovskaya, Yu.N. Makurin et al.: Doklady Phys. Chem. **391**, 187 (2003)
85. A.N. Enyashin, G. Seifert: Phys. Stat. Sol. (b) **242**, 1361 (2005)
86. A.N. Enyashin, V.V. Ivanovskaya, A.L. Ivanovskii: Mendeleev Commun. **14**, 94 (2004)
87. V.V. Ivanovskaya, A.N. Enyashin, A.A. Sofronov et al.: Solid State Comm. **126**, 489 (2003)
88. A.N. Enyashin, V.V. Ivanovskaya, Yu.N. Makurin et al.: Phys. Lett. A. **326**, 152 (2004)
89. A.N. Enyashin, V.V. Ivanovskaya, Yu.N. Makurin et al.: Chem. Phys. Lett. **392**, 555 (2004)
90. N. Grobert, T. Seeger, G. Seifert et al.: J. Ceramic Proc. Res. **4**, 1 (2003)
91. T. Seeger, Th. Köhler, Th. Frauenheim et al.: Chem. Commun. **34** (2002)
92. G. Bilalbegović: Phys. Rev. B **70**, 045407 (2004)

3

Spintronics: Transport Phenomena in Magnetic Nanostructures

Peter Zahn

Summary. Nanotechnology plays a decisive role in information technology. However the rapid increase (doubling of the Internet traffic every 6 months, of the wireless capacities every 9 months and of the magnetic information storage every 15 months) cannot be compensated by a simple downscaling of the semiconductor devices, as it was done in the past 30 years. To keep up with the demands, completely new devices have to be invented, operating on the nanoscale and exploit quantum effects. A very promising option is to use the spin of the electron in addition to its charge for information transmission and storage, i.e. going from the conventional electronics to spintronics. The foundations of this technique and the broadest application areas today, exploiting the giant magnetoresistance and the tunneling magnetoresistance are discussed from the experimental and theoretical point of view.

3.1 Introduction

The information technology revolution is based on an exponential rate of technological progress. For example, Internet traffic doubles every 6 months, wireless capacity doubles every 9 months, and magnetic information storage capacity doubles every 15 months. Moore's law which indicates that the performance of semiconductor devices doubles every 18 month has been valid for three decades. But, fundamental laws of physics limit the shrinkage of semiconductor components on which Moore's law is based, at least on current technologies. The continuation of the information technology revolution relies on new ideas for information storage and processing, leading to future applications. One option is to look for mechanisms that operate at the nanoscale and exploit quantum effects [1]. Nanotechnology covers a wide range of different technologies involved in the investigation, manipulation and control of matter on the very small scale, atom-by-atom and molecule-by-molecule. Such technology opens the possibility to develop materials and products with 'nanoscale' structures or to build devices and systems the same size as biological cells with highly desirable properties.

For a long time the charge of the electron was used to process and store information. But the electron has an additional degree of freedom – the spin. To exploit the spin for new information processing techniques and to integrate it into the traditional electronics, this is the vision of the emerging field of spin electronics or spintronics. Reviews with a broader scope than this lecture are given in [2–4]. Here the focus will be given to the effects of giant magnetoresistance and tunneling magnetoresistance.

In the field of metallic systems layered structures of magnetic and non-magnetic materials dominated the common interest. In these multilayers ferromagnetic layers are separated by non-magnetic spacer layers. The phenomenon of interlayer exchange coupling (IEC), discovered 1986 by Grünberg et al., favors one relative orientation of the magnetization direction of the ferromagnetic layers [5]. That is, forced by the exchange interaction mediated by the conduction electrons of the non-magnetic spacer layer, the moments of adjacent magnetic layers are aligned parallel or antiparallel in zero magnetic field. The sign and strength of the coupling are mainly determined by the material and the thickness of the non-magnetic spacer layer [6, 7].

Investigating the transport properties of these structures a new phenomenon, the effect of giant magnetoresistance (GMR) was found in 1988 [8,9]. This is a drastic change in the electrical resistivity under an external magnetic field. The magnetic field induces a change in the relative orientation of the magnetic layers. The GMR effect allows to turn the information of a two-state magnetic system (parallel or anti-parallel representing 0 or 1) into an electrical one, or in a more abstract sense, to translate spin information into charge current information. Secondly, a GMR device can easily detect the direction of a magnetic field. This opened a huge market for sensor applications. A comprehensive review of experimental results with GMR devices is given in [10] and an overview on theoretical models applied is given in [11].

Soon after the discovery of GMR, experiments have been carried out in which the non-magnetic metallic spacer was replaced by a non-magnetic insulator. In this geometry spin-polarized electrons tunnel from one ferromagnetic layer through an insulating barrier into the second ferromagnetic layer, and again a strong dependence of the resistance upon the relative orientation of the magnetization was found. The effect is called the tunneling magnetoresistance (TMR) and the device is called a magnetic tunnel junction (MTJ). In contrast to GMR systems, TMR systems exhibit a large voltage drop across the MTJ and operate with small electrical currents. The large technological interest on these systems has initiated a large number of experimental as well as theoretical investigations to elucidate the microscopic origin of the phenomena. Theoretical models to describe TMR in connection with experimental results are presented in [12].

The very promising field connecting spin electronics and classical semiconductor technology arises from the easy control of charge density in these devices by means of doping and gate electrodes. Recently suggested or demonstrated devices, like the Datta-Das transistor [13] and the spin-valve transistor

[14] will not be covered by this overview. New spin-based multifunctional devices merging electronics, magnetics, and photonics are spin light-emitting diodes, spin resonant tunneling devices, optical switches operating at terahertz frequency, modulators, encoders, decoders, and quantum bits for quantum computation and communication.

This lecture will be organized as follows. A formalism calculating the electronic structure and the transport coefficients of layered nanostructured materials without adjustable parameters in the framework of density functional theory will be presented here. The scheme will be applied to elucidate the microscopic origins of the interlayer exchange coupling phenomenon and the GMR effect in metallic multilayer films. Concerning the TMR effect the influence of the electrode material specifics and the electrode/barrier interface will be discussed at the end of the lecture.

3.2 Magnetism in Nanostructures

3.2.1 Magnetism in Reduced Dimensions

As the dimensions of the nanostructures are at least in one dimension of the order of a few atomic distances the influence of surfaces and interfaces becomes decisive for the material properties. Regarding spin electronics the evolution of magnetism is a key issue.

First consider, how magnetism evolves when the dimensionality of the system is decreased. The most important property of metallic surface atoms is the reduced coordination, the number of next nearest neighbors is smaller than in the bulk material. Applying a simple tight-binding picture and keeping the hopping matrix elements constant this results in a reduced band width of the electronic states. This in turn implies an increase in the local density of states, because the total number of electronic states is kept constant. The Stoner model gives a criterion for the occurrence of band magnetism by the product of the density of states (DOS) at the Fermi level and a local material dependent exchange integral [15] which is of the order of some tenth of an eV. This enhancement of the local DOS strengthen the tendency towards magnetism. Because the exchange integral increases going from 5d to 4d and 3d metals and the DOS is largest in the middle of the band, the strongest effect is expected in the middle of the 3d series. As can be seen from Tab. 3.1 the magnetic moment of the 3d metals is strongly enhanced at the surface with respect to the bulk materials.

A systematic theoretical study of the 4d metals in nanostructures of different dimensionality is shown in Fig. 3.1, adapted from Ref. [18]. All the elements are non-magnetic as a bulk material. The two-dimensional case (2D) represents a monolayer of the considered material on a non-magnetic Ag(001) substrate. In this case the coordination number is further reduced with respect to the surface and magnetism occurs for some of the materials. A further reduction of the coordination number is obtained in the one-dimensional case,

Table 3.1. Calculated magnetic moments of 3d metals: bulk vs. (001) surface [16,17]

μ_B	Cr	Fe	Co	Ni
$M_{surf}^{(100)}$	2.2	2.87	1.75	0.59
$M_{bulk}^{(100)}$	0.6	2.18	1.58	0.55

this corresponds to monoatomic wires suspended on a non-magnetic target, and the zero-dimensional case, where an adatom was considered on an otherwise perfect Ag surface. This causes an increase of the DOS at the Fermi level and a higher magnetic moment in most cases. It demonstrates that the dimensionality of the structures is a very sensitive parameter for the occurrence of magnetism. The technological exploitation of this dimensionality induced magnetism can not be anticipated at the moment.

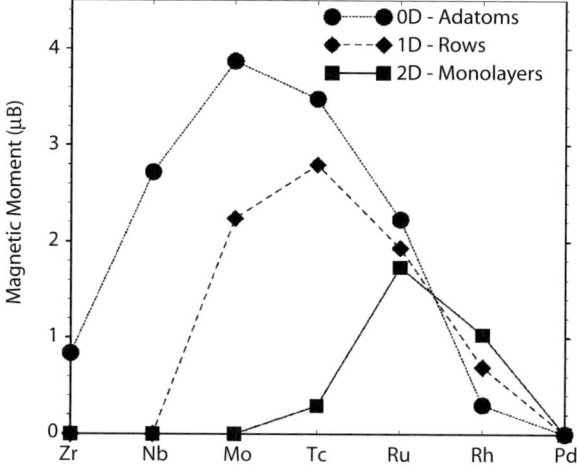

Fig. 3.1. Influence of the dimensionality on the magnetic moment profile. Shown are the results for adatoms (*circles*), infinite close-packed rows (*diamonds*), and monolayers (*squares*) of the 4d elements on the Ag(001) surface, from [18]

3.2.2 First Principals Calculational Scheme

In addition to model calculations ab-initio schemes are of increasing importance for the understanding of the microscopic magnetic behavior, because they are able to include the material specific properties.

Starting from density-functional theory the main task is to determine the electron density for a given effective potential. This contains the external potential, which defines the geometry of the system by the Coulomb field of the atomic nuclei.

Korringa [19], Kohn and Rostoker [20] (KKR method) invented one of the earliest, but up to now one of the most accurate electronic structure method, which is based on the multiple scattering method first derived by Rayleigh and others [21, 22]. It is a multiple scattering formalism to determine the Greens function and the eigenfunctions of a given potential by the superposition of partially scattered waves caused by scattering centers the potential is divided into. This allows for a separation of the properties of the local scattering centers and their geometrical arrangement. The electronic properties are derived from the one-particle Greens function (GF) of the considered system. In the following years the method was elaborated further to allow for self-consistent calculations of periodic crystals, and the full cell potential, to treat localized defects, to determine forces and lattice relaxations, and to treat surfaces, layered systems, point defects at surfaces, one-dimensional systems

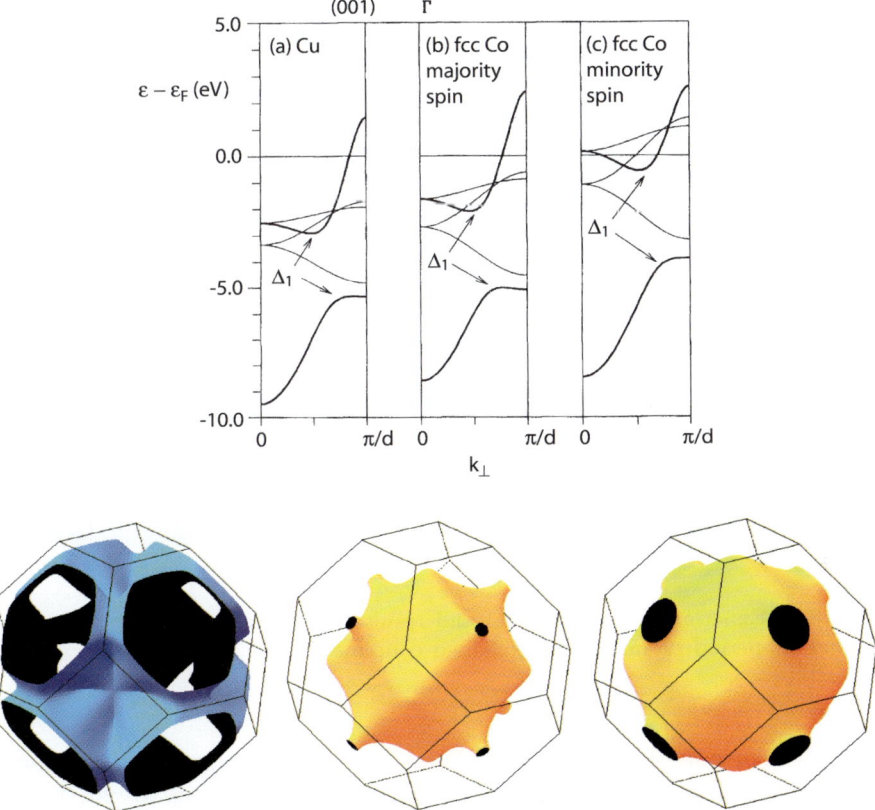

Fig. 3.2. Examples of self-consistent electronic structure calculation output: Band structure (*top*, from [7]) and Fermi surface (*bottom*) for Co minority electrons (largest Fermi surface contribution from 4th band), Co majority electrons and Cu electrons (from *left* to *right*)

and two-dimensional systems with open and confining boundary conditions in the third direction. A special summary of multiple scattering methods related to KKR is given in the proceedings of a MRS symposium [23] and a review of the recent conceptual improvements is given in [24].

The basic output of a self-consistent electronic structure calculation is the position dependent electron density. In addition, the energy bands E_k^σ and eigenstates Ψ_k^σ defined by the Kohn–Sham one-electron equation are obtained [25]. The index k is a shorthand notation for the wave vector $\boldsymbol{k}$ and the band index ν. σ denotes the spin directions 'up' for majority states and 'down' for minority states. Figure 3.2 shows in the upper part the band structure for Co along the ΓX direction of the fcc Brillouin zone. Due to the spin-splitting in the magnetic Co the bands for majority and minority electrons are shifted with respect to each other.

The Fermi velocities and the Fermi surface are of special interest in the following transport calculations. They are defined by the derivative of the energy bands and the set of k points with an eigenvalue of the energy E_k^σ equal to the Fermi level E_F, respectively. Examples for Fermi surfaces are given in Fig. 3.2 and can be found on the web [26]. The similarities of the Co majority band and the spin-degenerated Cu Fermi surface are obvious. In contrast, the Fermi surface of minority electrons points to differences of both metals.

3.2.3 Magnetic Interlayer Exchange Coupling

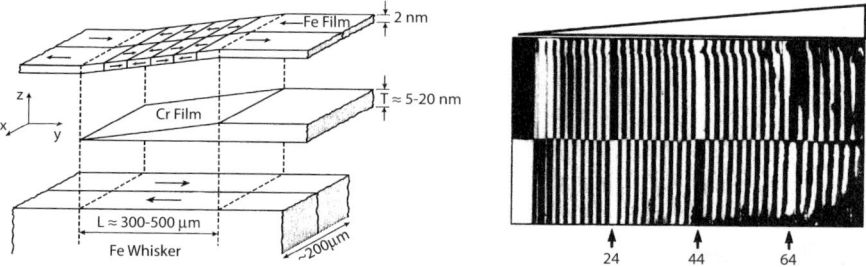

Fig. 3.3. Experimental setup (*left*) and result (*right*) to determine strength and oscillation period of interlayer exchange coupling of the Fe/Cr/Fe system, from [27, 28]

The phenomenon of magnetic interlayer exchange coupling can be found in metallic multilayered systems. The magnetic layers are separated by non-magnetic metal layers. An exchange coupling mediated by the electron gas of the non-magnetic layer causes an alignment of the moments of adjacent magnetic layers. Depending on the thickness of the non-magnetic interlayer a parallel or antiparallel alignement is prefered. This was demonstrated experimentally for the first time by Grünberg et al. in 1986. The strength of the coupling oscillates with the interlayer thickness, that means, the sign of the coupling strength changes periodically [6, 27, 29]. The black and white stripes in Fig. 3.3 mark the regions of parallel and antiparallel alignement of the bottom and top layer of a Fe/Cr/Fe trilayer. Due to the wedge shape of the Cr interlayer, regions with parallel and antiparallel alignement alternate in the direction of increasing Cr thickness.

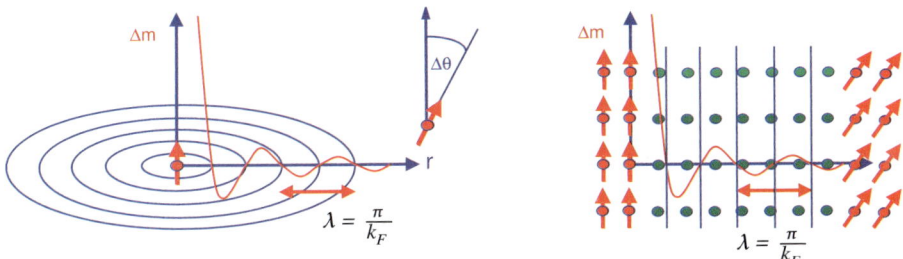

Fig. 3.4. Friedel oszillations of induced magnetization density in a nonmagnetic metal caused by a magnetic point defect (*left*) and a magnetic layer (*right*). A second magnetic entity with a magnetization tilted by an angle θ is sketched

In the limit of weak coupling, which is given for sufficient thick spacer materials, the oscillating behaviour can be described by an RKKY-like interaction (Ruderman, Kittel, Kasuya and Yoshida [30]). The magnetic layer forms a magnetic perturbation to the nonmagnetic spacer, which is screened by the electron gas, and in analog to the Friedel oscillations [31], a damped magnetization density wave will be induced. The adjacent magnetic layer interacts with the induced magnetization. The shape of this perturbation is sketched in Fig. 3.4. The decay of the induced magnetization density depends on the dimensionality of the electron gas and the perurbation. For the case of a point defect in a three-dimensional electron gas, shown in the left panel, the decay is proprotional $1/r^3$, with r the distance from the defect, and in the case of a planar perturbation, shown in the right panel, the decay is proprotional $1/z^2$, with z the distance from the magnetic plane. The period of the oscillation is defined by extremal callipers of the spacer material Fermi surface [32, 33] and is in very good agreement with the experimental findings for nearly all systems under consideration [34–36].

A general trend obtained from the RKKY models is that the strength of the coupling depends essentially on the degree of the matching of the energy bands at the magnet/nonmagnet interface [7]. This can be determined by ab initio calculations of the coupling energy [39–42]. Nevertheless, the coupling strength obtained from total-energy calculations for systems with ideal interfaces are typically one order of magnitude (or even more) larger than the experimental ones [43]. By including interface roughness and defects

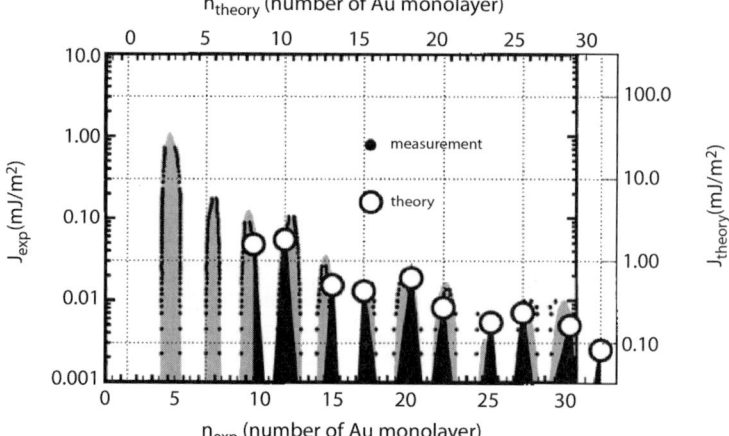

Fig. 3.5. Strength of interlayer exchange coupling in dependence on spacer thickness for Fe/Au/Fe: comparison of theory [37] and experiment [38]

at the interface to simulate the experimental situation more realistic, a very good agreement of theory and experiment could be achieved. Figure 3.5 shows a comparison for Fe/Au/Fe systems [37]. The position of the coupling maxima, the period and the absolute values of the coupling strength coincide. So, it can be summarized that the mechanism of the interlayer exchange coupling is rather extensively understood. The same physics dominates the transport phenomena found in these nanostructures, which are described in the following section.

3.3 Transport Phenomena

The theoretical concepts to describe diffusive and coherent transport, respectively, will be introduced in this section. The main part is devoted to the microscopic origins of GMR. The sensitivity of the TMR effect to the electronic structure of leads and insulator, as well as the geometry of the interfaces will be described at the end.

3.3.1 Transport Theory

To describe the transport properties of a solid is a challenging task for the theory. By definition the transport coefficients are related to a non-equilibrium situation which is not easily to handle. In the regime of linear response for weak external fields the description can be made by groundstate properties, which is founded by the fluctuation-dissipation theorem. This was demonstrated in particular for the case of electrical conductivity by Kubo [44].

Diffusive and Coherent Transport Regime

The conductance g characterizes the property of a sample to allow for an electrical current when a voltage is supplied. In the regime of linear response the flowing current I is proportional to the applied voltage U

$$g = \frac{I}{U} \; . \tag{3.1}$$

In dependence on the scattering properties of the sample the transport is called diffusive or coherent. This is sketched in Fig. 3.6. In case that the mean free path of the electrons is much shorter than the length of the conductor, many scattering processes occur in the sample and the phase coherence of the incoming and outgoing states in not conserved. In this case the conductance can be characterized by a material specific conductivity σ, the length L and the cross section A of the sample

$$g = \sigma \frac{A}{L} \; . \tag{3.2}$$

In the opposite case, Fig. 3.6, right panel, the mean free path is much longer than the sample dimensions and no scattering events at randomly distributed defects occurs. In this case the conductance depends in a non-trivial way on the sample dimensions.

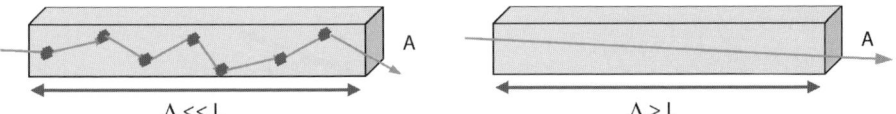

Fig. 3.6. Limits of transport measurements: diffusive transport (*left*) and ballistic transport (*right*). The cross section of the sample is A and the length along the current direction is L

For the first case of diffusive transport and a small scattering rate, that means low defect concentration or weak scattering, the quasi-classical Boltzmann approach is appropriate. The main ideas will be given below. The Kubo formalism can even treat systems with strong disorder or scattering and is equivalent to the Boltzmann approach in the weak scattering limit. The limit of strong scattering is out of the scope of this lecture. Both approaches were successfully applied to describe the residual resistivity of metals, multilayers and the GMR effect [45–49].

Nanowires [50, 51], tunneling junctions [52, 53], metallic multilayer pillars showing current-perpendicular-to-plane GMR [54], or molecular junctions [55] are samples much shorter than the electron mean free path. In these cases, the Landauer-Büttiker formalism [56, 57] is well suited to calculate the conductance by means of transmission coefficients. Methods based on layer-KKR [58], tight-binding formulations [59], or a transfer matrix scheme [60] were applied to describe the transport in tunnel junctions.

Boltzmann Theory

The Boltzmann theory is based on a classical distribution function $f_k(\boldsymbol{r}, t)$ which gives the number of carriers in quantum-mechanical state k, and can depend on the position in real space $\boldsymbol{r}$, too. k denotes the wave vector $\boldsymbol{k}$, the band index ν, and for magnetic systems the spin σ. In the following derivations the spin is neglected for the sake of simplicity and an explicit dependence of the distribution function on time and magnetic field is excluded. The real space dependence vanishes due to the restriction to homogenous systems. In the steady state the total rate of change has to vanish, and from the conservation of phase space volume a master equation for the distribution function is derived

$$\dot{\boldsymbol{k}}\frac{\partial f_k}{\partial E_k}\frac{\partial E_k}{\partial \boldsymbol{k}} - \frac{\partial f_k}{\partial t}\bigg|_{scatt} = 0 \qquad \text{with} \qquad \boldsymbol{v}_k = \frac{\partial E_k}{\partial \boldsymbol{k}} \ . \tag{3.3}$$

The drift term is determined by the external electric field $\boldsymbol{E}$ with e the electron charge $e = -|e|$ and $\dot{\boldsymbol{k}} = e\boldsymbol{E}$. The second term in (3.3) describes the change of carriers in state k due to scattering, and is determined by the microscopic transition probability $P_{kk'}$ via (3.4). These scattering processes can be caused, e.g. by lattice defects, imperfections or thermally activated quasi particles. This is shown for the case of a point defect in Fig. 3.7.

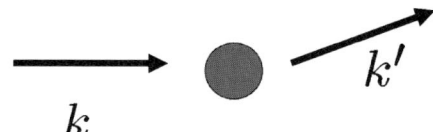

Fig. 3.7. Source of resistance: scattering of a state k into a state k' at a point defect

The microscopic transition probability $P_{kk'}$ is given by Fermi's Golden rule

$$P_{kk'} = 2\pi c N \left|T_{kk'}\right|^2 \delta\left(E_k - E_{k'}\right) \ . \tag{3.4}$$

$T_{kk'}$ describes the scattering at one impurity. cN is the total number of impurities in the sample with c the relative concentration of defects. In the frame work of a KKR Greens function formalism the transition matrix elements $T_{kk'}$ can be calculated without adjustable parameter using the self consistently determined potentials of isolated defects in an otherwise perfect matrix [61,62].

Exploiting the microscopic reversibility $P_{kk'} = P_{k'k}$, considering energy conserving scattering processes only, and introducing the small deviation $g_k = f_k - \mathring{f}_k$ from the equilibrium distribution function $\mathring{f}_k$, the change of the distribution function due to scattering is given by

$$\frac{\partial f_k}{\partial t}\bigg|_{scatt} = \sum_{k'} P_{kk'}\left(\tilde{f}_{k'} - \tilde{f}_k\right) \ . \tag{3.5}$$

In the limit of linear response the proportionality of g_k and the external field $\boldsymbol{E}$ is given by the vector mean free path $\boldsymbol{\Lambda}_k$ using the ansatz

$$\tilde{f}_k = -e \frac{\partial \mathring{f}_k}{\partial E} \boldsymbol{\Lambda}_k \boldsymbol{E} \ . \tag{3.6}$$

Neglecting higher order terms in $\boldsymbol{E}$ and using the relaxation time $[\tau_k]^{-1} = \sum_{k'} P_{kk'}$ one obtains with (3.3) and (3.5) the linearized Boltzmann equation

$$\boldsymbol{\Lambda}_k = \tau_k \left[\boldsymbol{v}_k + \sum_{k'} P_{kk'} \boldsymbol{\Lambda}_{k'} \right] \ . \tag{3.7}$$

The first term on the r.h.s describes the means free path in relaxation time approximation known from textbooks. The second term on the r.h.s. is the so-called scattering-in term which counts the scattering events from states k' back to the considered state k. It causes the vertex corrections in the expansion of the mean free path. In the limit of zero temperature the integration of (3.7) has to be done for the states on the anisotropic Fermi surface of the system under consideration. For magnetic systems the spin variable σ has to be considered, which can easily be included formally by $k = (\boldsymbol{k}, \nu, \sigma)$.

A simplified model to calculate the relaxation times was proposed by the author and successfully applied to the transport properties of metallic multi-layers and the effect of giant magnetoresistance [63].To focus on the influence of the superlattice wave functions, the details of the scattering potential are neglected and δ-scatterers with a spin-dependent scattering strength t^σ at lattice sites μ are assumed. Consequently, the spin-dependent relaxation time in Born approximation becomes [63]

$$[\tau_k^\sigma (\mu)]^{-1} = 2\pi c \left| \mathring{\Psi}_k^\sigma (\mu) \right|^2 n^\sigma (\mu, E_\mathrm{F}) (t^\sigma)^2 + \tau^{-1} \ . \tag{3.8}$$

The relaxation time $\tau_k^\sigma (\mu)$ of state k is inverse proportional to the probability amplitude $\left| \mathring{\Psi}_k^\sigma (\mu) \right|^2$ and the local density of states $n^\sigma (\mu, E_\mathrm{F})$ at Fermi level at the position μ of the defect. To avoid short circuit effects due to states with a tiny probability amplitude at the impurity position a constant inverse relaxation time τ^{-1} is added, which represents all other scattering processes.

Residual Resistivity

To calculate the residual resistivity, that is the resistivity at zero temperature caused by scattering at defects only, the current density is expressed by the deviation g_k of the distribution function

$$\boldsymbol{j} = -\frac{e}{V} \sum_k \boldsymbol{v}_k \tilde{f}_k \ . \tag{3.9}$$

The contribution of the equilibrium occupation function $\overset{\circ}{f}$ vanishes. In the limit of zero temperature and linear response the non-zero contributions of g_k are restricted to the Fermi surface and the current density can be expressed by a Fermi surface integral, using the ansatz in (3.6)

$$\sigma^{ij} = \frac{e^2}{(2\pi)^3} \oint_{E_k=E_F} \frac{\mathrm{d}S}{v_k} v_k^i \Lambda_k^j \; , \tag{3.10}$$

with i and j the Cartesian coordinates and $\mathrm{d}S$ a differential Fermi surface element. In non-magnetic systems a factor 2 has to be included to account for the spin-degeneracy.

Applying the relaxation time approximation for the evaluation of the vector mean free path, the Fermi surface integral for the conductivity contains the tensor of the Fermi velocities scaled by the state dependent relaxation time

$$\sigma^{ij} = \frac{e^2}{(2\pi)^3} \int_{E_k=E_F} \frac{\mathrm{d}S}{v_k} v_k^i v_k^j \tau_k \; . \tag{3.11}$$

The Fermi surface integral contains information about the band structure of the unperturbed system only, and τ_k contains the information about the scattering properties.

In magnetic systems with negligible spin-flip scattering both spin channels contribute to the current in parallel applying Motts two-current-model [64]. So, the conductance is split in majority and minority contributions

$$\sigma = \sigma^\uparrow + \sigma^\downarrow \; . \tag{3.12}$$

Without spin-orbit coupling and without magnetic fields the conductivity tensor is diagonal. In layered single crystalline metallic systems the conductivity tensor contains 2 different entries. That is for the current in direction perpendicular to the planes of the layers (current-perpendicular-to-plane, CPP), that is parallel to the crystal c-axis, the σ_{CPP}, and for the current in direction of the plane of the layers (current-in-plane, CIP) the σ_{CIP} conductivity

$$\sigma_{\mathrm{CPP}} = \sigma^{zz} = \sigma_\perp \quad \text{and}$$
$$\sigma_{\mathrm{CIP}} = \sigma^{xx} = \sigma^{yy} = \sigma_\parallel \; . \tag{3.13}$$

Landauer Theory

For mesoscopic systems where the system size plays a crucial role for transport properties the Landauer-Büttiker theory is an established method [56,65–67]. The considered device contains a scattering region which is connected to electron reservoirs by ideal leads which feed the current into and take it out. The measurement geometry is illustrated in Fig. 3.8 and determines the assumption to be made in the application of the formalism. A two-terminal device is

Fig. 3.8. Geometry for conductance measurements: A scattering region is connected by ideal leads to electron reservoirs (baths), which determine the voltage drop at the junction, **I** denotes the incident flux, **T** the transmitted, and **R** the reflected one

characterized by two reservoirs and current as well as voltage are measured at the same leads. It is assumed that the elements of the transmission matrix T are small compared to unity, which describes a tunnel junction. In the case of zero temperature and within linear response the Landauer conductance becomes

$$g = \frac{e^2}{h} \sum_{\substack{k_{\mathrm{L}}, k_{\mathrm{R}} \\ E_{k_{\mathrm{L}}} = E_{k_{\mathrm{R}}} = E_{\mathrm{F}}}} T_{k_{\mathrm{L}} k_{\mathrm{R}}} \ , \qquad (3.14)$$

with $T_{k_{\mathrm{L}}, k_{\mathrm{R}}}$ the quantum mechanical probability for a state k_{L} incident from the left reservoir to be transmitted into state k_{R} of the right hand side reservoir. k_{L} and k_{R} are the quantum numbers of the eigenstates in the left and right hand side electrode, respectively, which are normalized to carry a unit current. This formula was empirically derived for one-dimensional systems. Fisher and Lee have later shown that (3.14) can be extended to systems of higher dimensions using the transmission matrix t connecting the incident flux in the various channels k_{L} to the outgoing flux in the channels k_{R} on the other side [68]

$$g = \frac{e^2}{h} \operatorname{Tr} \left(t^+ t \right) \ . \qquad (3.15)$$

For planar tunnel junctions with an in-plane translational invariance and magnetic electrodes $k_{\mathrm{L,R}}$ is a shorthand notation for $(\boldsymbol{k}_\parallel, \nu, \sigma)$, the in-plane wave vector $\boldsymbol{k}_\parallel$, band index ν, and spin σ.

The transmission as introduced by Landauer can be computed using a Kubo formalism in the zero bias limit by means of the Green's function of the semi-infinite system [69]. The conductance g of a planar junction is obtained by a 2-dimensional integration over the surface Brillouin zone [70]

$$g = \sum_\sigma \mathrm{d}^2 \mathbf{k}_\parallel T_{\mathbf{k}_\parallel}(E_{\mathrm{F}}) \ , \qquad (3.16)$$

$$T_{\mathbf{k}_\parallel}(E) = tr \left[J_{\mathrm{L}}(E) G_{\mathrm{LR}}(\mathbf{k}_\parallel, E) J_{\mathrm{R}}(E) G_{\mathrm{RL}}(\mathbf{k}_\parallel, E) \right] \ . \qquad (3.17)$$

The planes L and R are situated on both sides of the barrier in the unperturbed electrode regions. $J_{\mathrm{L,R}}(E)$ are the current operator matrices and $G_{\mathrm{LR}}(\mathbf{k}_\parallel, E)$

are the Green's function elements connecting both sides of the junction. The transmission $T_{\mathbf{k}_\parallel}(E)$ contains the transmission coefficients T_m of the states fed by the reservoirs as introduced by Büttiker [57] as eigenvalues of the matrix.

3.3.2 Giant Magnetoresistance

Basics

The giant magnetoresistance effect (GMR) occurs in metallic multilayer systems where magnetic layers are separated by non-magnetic interlayers. The change of the relative orientation of the magnetization directions of adjacent magnetic layers is accompanied by a drastic change in the electrical resistivity. This reorientation of the magnetization is mostly driven by an external magnetic field.

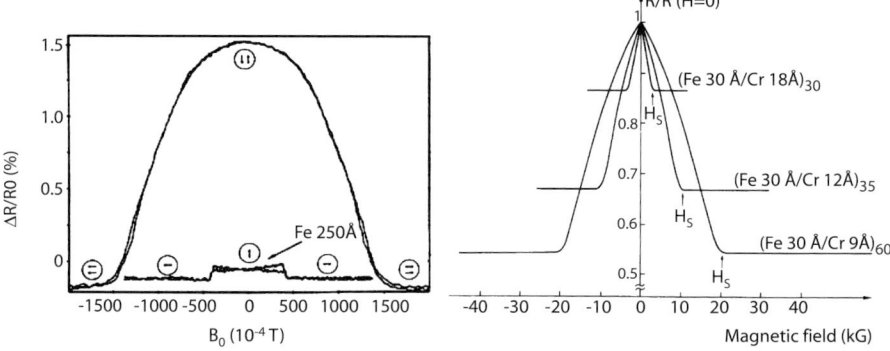

Fig. 3.9. First experimental proof of GMR in Fe/Cr/Fe: trilayer result from [9] (*left*), multilayer result from [8] (*right*), 30, 35 and 60 double layers were deposited in the stack, resp.

To quantify the change in resistivity the GMR ratio is introduced

$$\mathrm{GMR} = \frac{g^\mathrm{P} - g^\mathrm{AP}}{g^\mathrm{AP}} \ , \tag{3.18}$$

with g^P and g^AP the conductivities for the corresponding parallel (P) and antiparallel (AP) magnetic configuration, which are the inverse of the resistivities R^P and R^AP, respectively. In most cases a positive GMR ratio is obtained, see the first experimental evidence from 1988 in Fig. 3.9. Investigating the dependence of the resistivity change on the relative angle of the layer magnetizations a cosine dependence was found. An example is given in Fig. 3.10. Higher order terms in cosine of the angle are much smaller and allow for an easy application in angular and position sensing devices.

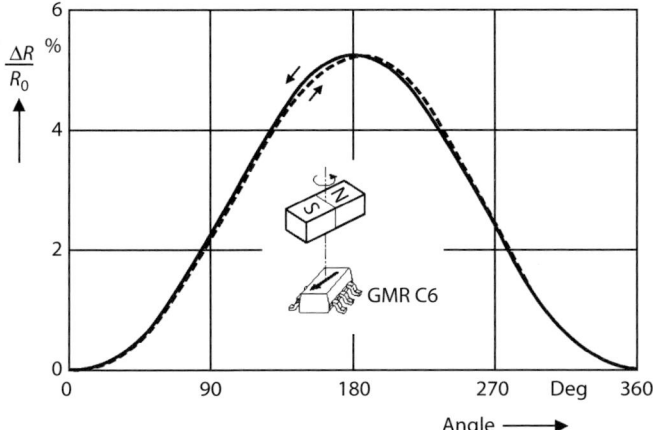

Fig. 3.10. Change of resistivity vs. angle between the layer magnetizations of a GMR sensor element, the insets shows the commercial sensor GMR-C6 of Infineon Inc. together with a magnetic dipole causing the field variations, from [71]

The main reason for the occurrence of GMR is the matching of the spin split electronic states in the magnetic layers and the spin-degenerated states in the non-magnetic interlayers. This is very pronounced in the systems Co/Cu and Fe/Cr, the working horses of spintronics. The different matching of the electronic states for a parallel and antiparallel alignment of moments causes a hybridization of states with different group velocities. This results in a drastic change of Fermi velocities of the states in the whole stack [46,72]. This effect is called the intrinsic GMR, because it is exclusively caused by changes of the coherent electronic structure of the multilayer. In experimental and theoretical investigations the importance of defects in the layers and especially at the interfaces was pointed out [73–77]. An overview on theoretical methods describing GMR is given in [11]. A detailed analysis of the influence of coherent electronic structure and scattering originated by defects will be given in the following.

Microscopic Origin

Despite other geometries and device layouts, GMR occurs for systems with a magnetic groundstate characterized by antiparallel orientation of the magnetic moments in adjacent magnetic layers caused by the effect of interlayer exchange coupling. Accordingly, we have chosen for the theoretical investigations a multilayer geometry in the so-called first antiferromagnetic maximum of interlayer exchange coupling consisting of 9 monolayer Co separated by 7 monolayer Cu, denoted as Co_9Cu_7. This case allows the description of interlayer exchange coupling and transport phenomena on equal footing.

The calculated antiferromagnetic maximum at a Cu thickness of 7 monolayer is in excellent agreement with experimental results [35, 78, 79] and other calculations [80].

The structure of the superlattice was assumed to be a fcc lattice with a lattice constant of 6.76 Bohr radii. Lattice distortions and mismatches are neglected. This is justified by structural investigations where the mismatch between Co and Cu was found to be small and epitaxial layer-by-layer growth was obtained under certain conditions [81].

An important ingredient to the microscopic understanding of the conductivity is the layerwise decomposed density of states (LDOS). The LDOS are calculated from the diagonal part of the spin-dependent one-particle Green function of the multilayer system. They can be resolved by means of the spectral representation of the Greens function into a superposition of probability

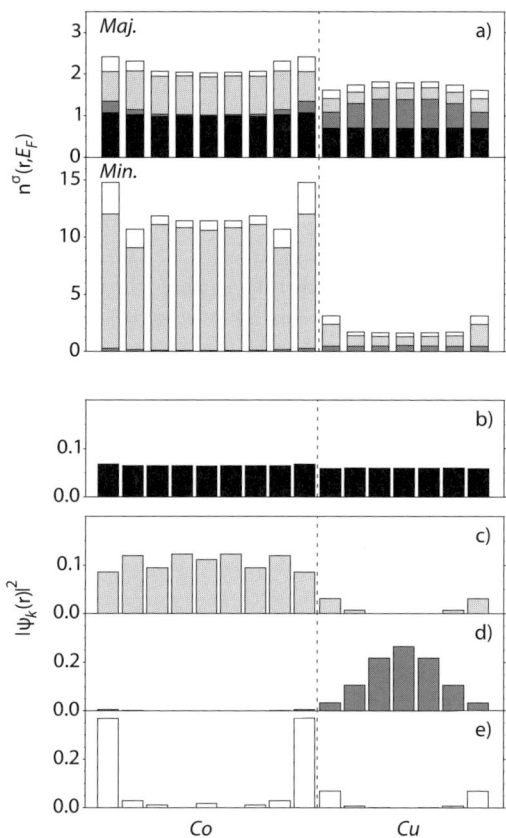

Fig. 3.11. Co_9Cu_7 in P configuration: spin-dependent local density of states at E_F in atomic units (**a**), probability amplitude of an extended majority state (**b**) and of different quantum well and interface states of minority spin (**c-e**) [63]

amplitudes of all eigenstates at energy E. The LDOS at E_F of the considered Co/Cu system for the P configuration is shown in Fig. 3.11(a) and is nearly the same at all monolayers in the majority channel [63]. The minority electrons are characterized by a very inhomogeneous profile. The LDOS in the Co layers are much higher than in the Cu layers. The largest values are obtained for the Co layers at the interface. This is a general behavior independent on Co or Cu layer thicknesses. This profile can be easily interpreted by means of probability amplitudes of the eigenstates. Due to a smooth potential profile most of the eigenstates in the majority bands are extended with a continuous probability amplitude for all layers, see Fig. 3.11(b). In contrast, the minority electrons are governed by a multi-well potential with a periodicity perpendicular to the layers (z-direction). For this reason quantum well states appear which are well localized perpendicular to the layers but extended in plane. Besides quantum well states with a localization in the center of the Co and Cu layers, Fig. 3.11(c,d), pronounced interface states in the Co layer are formed as shown in Fig. 3.11(e). The interface states can be understood in terms of resonant scattering and compares to the virtual bound state of a Co impurity in a Cu matrix [82].

In the AP configuration both spin channels are dominated by quantum well states. On average, these Bloch states are less extended than states in the majority band and less localized than in the minority band. The localization of the eigenstates determine the propagation of the electrons which has a direct influence on the transport properties.

To elucidate the influence of the quantum confinement of the eigenstates, the scattering properties of Cu impurities in the Co layers and Co impurities in the Cu layer are investigated. The eigenstates show strong quantum confinement due to the superlattice potential. That is, the probability amplitude

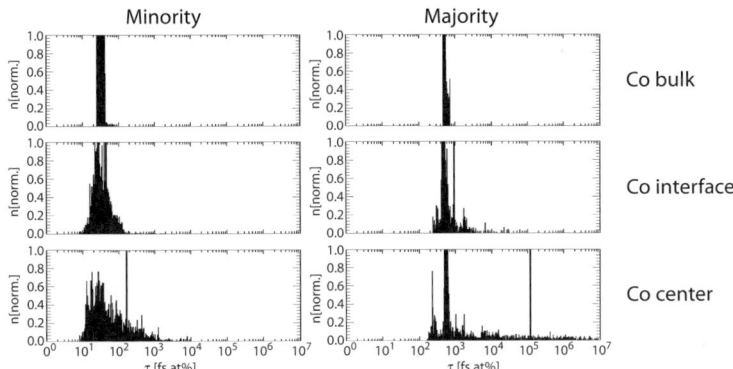

Fig. 3.12. Distribution of spin-dependent relaxation times caused by Cu impurities in Co bulk and at the Co interface and center sites of a $Co_9/Cu_7(001)$ multilayer in parallel magnetic configuration, the vertical lines mark the averaged relaxation time $\langle \tau_k \rangle_{E_F}$ [83]

is modulated by the layered structure and can even tend to zero at particular sites of the supercell [63,84]. As a consequence, all Bloch states with a nearly zero probability amplitude at the impurity site undergo a weak scattering and cause extremely large relaxation times. The anisotropy of the scattering can be characterized much more vivid by the spin and state dependent relaxation times, rather than by the resulting vectors of mean free paths including the vertex correction effects. The state dependent relaxation times are distributed over several orders of magnitude, especially for defects inside the metallic layers as shown in Fig. 3.12 by the probability $n[norm.]$ to find a state with a given logarithm of the relaxation time. This is a new effect that was never obtained in bulk systems and is related to the modulation of the Bloch states due to the layered structure [63].

Since the relaxation time in first Born approximation according to (3.8) is proportional to the spin-dependent LDOS at the impurity site, a strong position dependence of the conductivities is expected. Because of the large enhancement of LDOS at the Co interface layer impurities in these positions will be the most effective scatterers in comparison to all other positions. Furthermore, from the knowledge of spectral weights and probability amplitudes, it can be concluded that Co quantum well states are strongly scattered by defects at the interface position.

The results for GMR as a function of the impurity position in Co/Cu multilayer are shown in Fig. 3.13. The impurity position denotes an atomic layer, where the defects of the opposite species (Co or Cu) are introduced during the growth of the multilayer. This is called a δ-layer because the concentration of defects is rather small [77]. To simulate all other than the dopants in the δ-layer, defects at the interfaces were assumed, because structural investigations of Co/Cu multilayers on an atomic scale [86, 87] gave evidence that most of the structural imperfections appear next to the interfaces.Comparing the trend of GMR a surprisingly good agreement with the experiment (inset) is obtained, also concerning the trends of different dopant materials [85]. One has to mention, that the calculated values are two orders of magnitude larger than the experimental ones. The reason is the restriction to substitutional point defects. In addition to these much more scattering mechanisms are active in real samples. Assuming self-averaging the results could be corrected towards the experimental ones by an additional spin- and state-independent relaxation time τ [85].

In conclusion, the self-consistent calculation of the scattering properties and the treatment of the Boltzmann transport equation including vertex corrections provide a powerful tool for a comprehensive theoretical description and a helpful insight into the microscopic processes of GMR. The experimentally found trends concerning the doping with various materials at different positions in the magnetic multilayer could be reproduced for many different systems which means that spin-dependent impurity scattering is the most important source of GMR. The theoretical results show furthermore that interface scattering caused by intermixing plays a crucial role and has to be taken

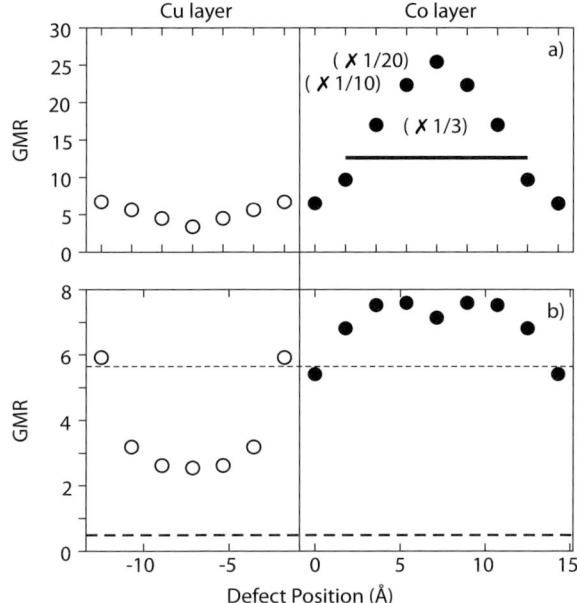

Fig. 3.13. Dependence of the GMR in $Co_9Cu_7(001)$ multilayer on the position of Co impurities in the Cu layer and Cu impurities in the Co layer: a) Assuming scattering at the inserted δ-layer only, b) Assuming δ-layer scattering and interface scattering with equal weights. The thick dashed line indicates the intrinsic GMR and the thin dashed line the GMR value caused by interface scattering only, from [85], the inset shows the experimental results by Marrows et al. [77]

into account in any system under consideration. Selective doping of the multilayer with impurities in specific positions causes variations of GMR which could be well understood by the modulation of spin-dependent scattering due to quantum confinement in the layered system and by the spin anisotropy of the scattering. The scale of discrepancies between calculations and experiments has to be considered as a measure of the importance of additional scattering processes.

Applications

GMR systems dominate nowadays widely the hard disk reading sensor technology, since the stray field of a hard disk causes a magnetization rotation in highly sensitive layers. The main advantage of GMR systems with respect to systems exploiting the anisotropic magnetoresistance is the larger signal amplitude of the resistivity change and the high potential for sensor miniaturization. Figure 3.14 illustrates the size development of the read heads driven by the exponentially increasing areal density of magnetic storage. The areal density has been increasing at a compound growth rate of 60% per year since 1991

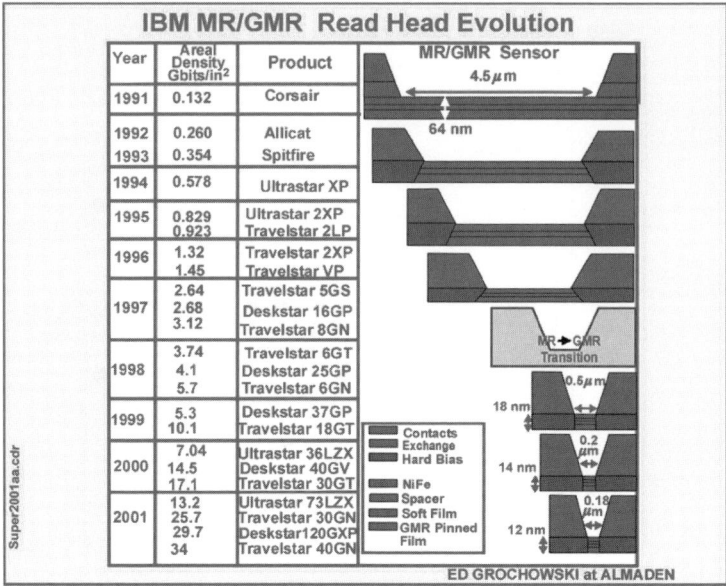

Fig. 3.14. Evolution of IBM hard disk areal density and dimension of read heads [88]

and 100% since 1998. The substitution of the GMR reading heads by the TMR technology due to further shrinking device dimensions is under development.

Another broad field of applications for the GMR devices is the sensor technology, despite the fact, that the hard disk reading head is a special type of magnetic field sensor. Aligning the magnetization from the AP configuration to the P one by an external field the nearly perfect cosine output signal of the GMR device allows for an easy evaluation, see Fig. 3.10. GMR sensors are commercially available for angular and linear movement, as well as pressure measurement since 1997. These sensors are ideal for all kinds of non-contact position registration (e.g. distances, speed, angular speed, sense of rotation) and for non-contact measurements of electrical currents and powers. For sensor applications, especially in the automotive field, technological aspects, like temperature stability, longterm reproducibility and reversibility are challenging tasks.

3.3.3 Tunneling Magnetoresistance

Basics

A large number of experimental and theoretical investigations to elucidate the microscopic origin of the phenomenon was initiated by the rediscovery of tunneling magnetoresistance (TMR) in metal/insulator/metal junctions [89,91] in 1995. New preparation techniques allowed for higher quality samples,

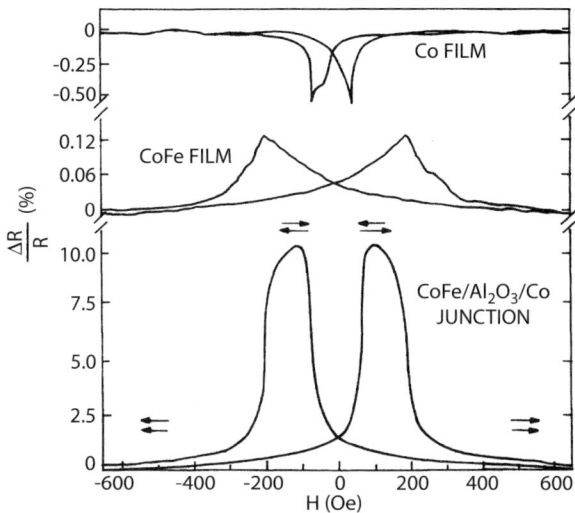

Fig. 3.15. Experimental results of remarkably high TMR values in CoFe/Al₂O₃/Co junctions: relative change of the resistance as function of magnetic field in the film plane, also shown is the variation in the CoFe and Co film resistance, from [89]

especially concerning the interface roughness, thickness and composition of the barrier. An example of the obtained magnetoresistance is shown in Fig. 3.15.

The TMR ratio is the normalized conductance difference for a parallel (P) and antiparallel (AP) alignment of the magnetic moments in the electrodes. Depending on the denominator a so-called pessimistic definition $\left(g^{\mathrm{P}} - g^{\mathrm{AP}}\right)/g^{\mathrm{P}}$, which gives smaller values for a positive TMR ratio, and a optimistic definition $\left(g^{\mathrm{P}} - g^{\mathrm{AP}}\right)/g^{\mathrm{AP}}$ is in use.

Tedrow and Meservey determined for the first time the spin polarization of a tunnel current injected by a ferromagnetic electrode using a superconducting counter electrode [92, 93]. To model the magnetoresistance results of a Fe/Ge/Co junction Julliere derived a simple formula using the spin polarization of the density of states at the Fermi level of the electrodes 1 and 2 $P_{1,2} = \left(n_{1,2}^{\uparrow}(E_{\mathrm{F}}) - n_{1,2}^{\downarrow}(E_{\mathrm{F}})\right)/\left(n_{1,2}^{\uparrow}(E_{\mathrm{F}}) + n_{1,2}^{\downarrow}(E_{\mathrm{F}})\right)$ [94] to express the optimistic TMR ratio as

$$\mathrm{TMR} = \frac{2P_1 P_2}{1 - P_1 P_2} \ . \tag{3.19}$$

This formula allows for a rough estimation of the effect by the experimentally determined polarizations of different electrode materials [95], but the properties of the barrier are neglected at all. The differences in the transmission probability for different states in the electrodes are washed out. Julliere's approximation is applicable in the dirty limit, where non-coherent scattering across the barrier dominates. Up to very recently, this estimate was to be considered as an upper limit for the TMR ratio for a given pair of electrodes and was supported by the experimental results. With the higher quality of the samples, much higher values than predicted by this formula have been achieved in recent experiments [52, 53, 96]. A second weak point of Julliere's argumentation is the meaning of the polarization, whether the surface, interface or bulk spin polarization of the density of states has to be taken. The strong influence of the spin polarization of the current or more general the electrode/barrier interface was verified in experiments [97–99]. Recent experiments point to the importance of the crystalinity of the barrier and the quality of the interfaces to obtain a large TMR ratio.

Microscopic Origin

A first theoretical description of the TMR in the framework of band theory was given by Slonczewski [100]. Expanding this model, the crucial influence of a non-magnetic layer at the interfaces in connection with disorder was confirmed [101].

For completely symmetric junctions it was shown, that resonant tunneling processes occur due to the large overlap of interface resonance states which are coupled to propagating states in the electrodes [90,102]. The total contribution of these resonances is not influenced by the additional occurrence of inelastic scattering [103]. In real samples these resonances are suppressed by disorder, which destroys the perfect symmetry.

The symmetry with respect to the interface of the states of the electrodes and inside the barrier mainly determines the transmission probability. In systems which are not translationally invariant electronic states with complex wave vectors are necessary, because Bloch states are not sufficient to make

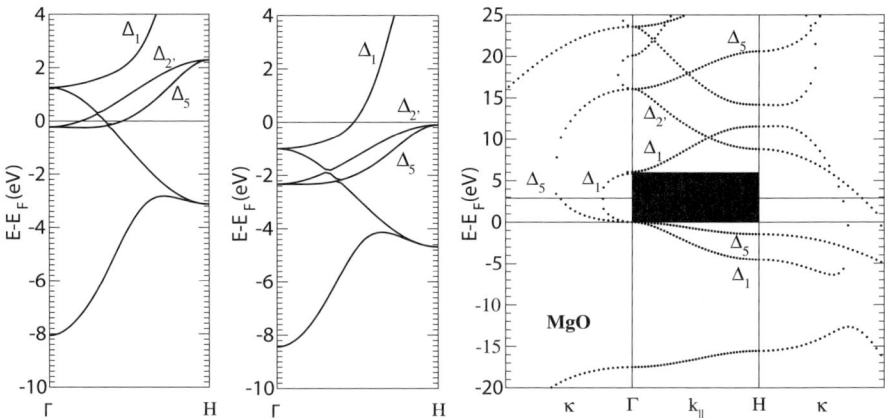

Fig. 3.16. Band structure and symmetry character of eigenstate for Fe electrode and MgO barrier material on the symmetry line Δ in reciprocal space, for MgO the states with complex wave vectors are given in addition

up a matched total wave function of the system [104, 105]. So the symmetry character and decay length of states with complex wave vector of the barrier material are essential for the tunneling probability derived from the transmission coefficient. The real bands of MgO along the Δ line ΓX are shown in Fig. 3.16, right panel, along with the corresponding eigenstates with complex wave numbers at the symmetry points Γ and X. In the band gap states different by symmetry and decay length appear.

Transmission coefficients for tunnel junctions in the scanning tunneling microscopy geometry including the electronic structure of both the electrodes and the vacuum barrier where first published on an ab initio level using a layer-KKR approach [58, 106]. Extending this approach to Fe/MgO/Fe systems the transmission as a function of $k_{\parallel}$, the conductance and the TMR in the limit of zero bias could be obtained [90]. These results are in very good agreement to TB calculations by Mathon et al.[107] and the corresponding results presented here.

Within the framework of a tight-binding model it was shown that impurities within the barrier cause defect states in the gap, increase conductance and decrease polarization [59]. This quenches the tunneling magnetoresistance.

Applying an external bias voltage the TMR effect decreases to half the maximum value at typical voltages of some tenth of a Volt. This is ascribed to the lowered effective spin polarization of the states feeding the current. With an external bias states from an energy window below or above the Fermi level are incorporated in transport. A second phenomenon was described as the 'zero bias anomaly' which describes a rapid decrease of TMR ratio within a voltage range of up to about 0.1 eV. Levy and Zhang have shown, that magnon scattering at the electrode/barrier interfaces can be one reason [108]. The inversion of the sign of tunneling magnetoresistance by an applied bias voltage was discussed in terms of energy dependent density of states and spin polarization of the electrodes [109, 110]. The absence of the TMR decrease with bias in spin-polarized vacuum tunneling with Co pointed to the importance of the interface scattering properties [111]. Combining Co electrodes and a MgO barrier a huge TMR ratio was predicted by calculations [112]. The bias dependence of the tunnel current of planar junctions was first discussed with ab initio methods in [113].

In the following ab initio calculations on the dependence of the transmission and the resulting TMR ratio on the structure of metal/insulator interfaces and the applied bias voltage will be discussed. This will be done for the Fe/MgO/Fe system. This system can be crown epitaxially, which ensures a large proximity of the experimental samples and the theoretical assumed structure.

Detailed structure investigations of the interfaces revealed the formation of a mixed FeO layer at the bottom interface with a 60% occupation of the O sites under certain growth conditions [114, 115]. Since the wetting of the materials differs, changes at the top interface are most probable, but detailed structural data are still missing.

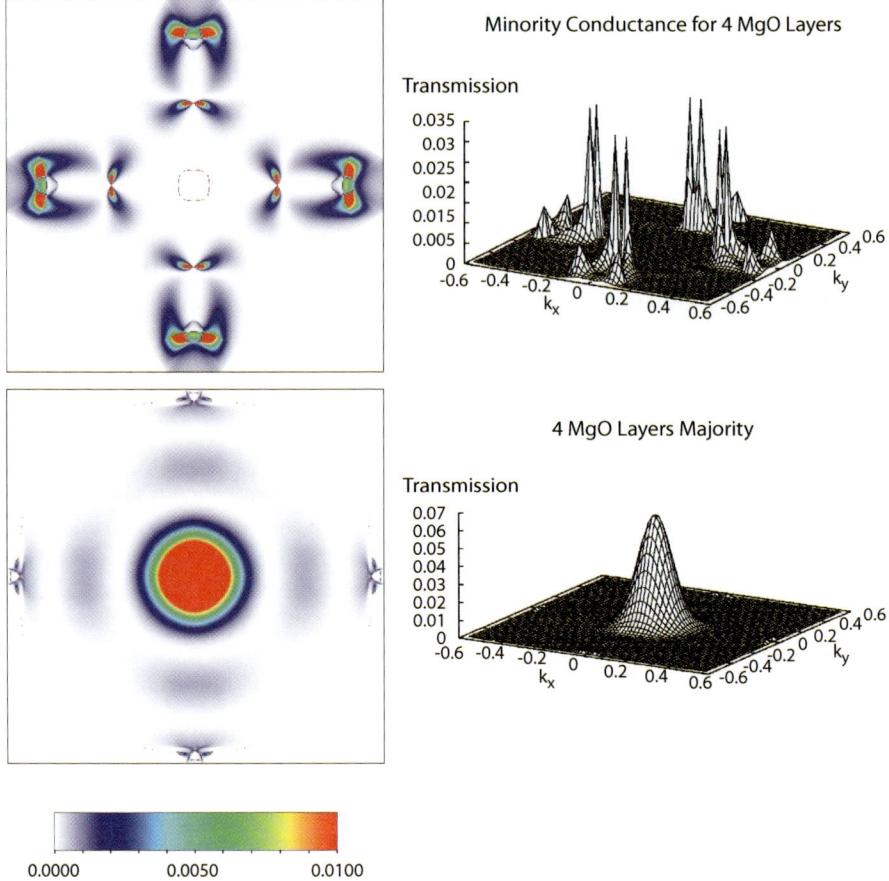

Fig. 3.17. Fe/4MgO/Fe junctions: Transmission probability in dependence on the in-plane wave vector $\mathbf{k}_\parallel$ for the minority (*upper row*) and the majority (*lower row*) electrons, the right column show results of Butler et al. [90] for comparison

To obtain a first estimate which states are most important for the tunnel current a model system with a single parabolic band will be exploited. The system is translationally invariant in x and y direction and the barrier has a height V and a thickness d along the z direction. Considering an incoming electron with an in-plane wavevector $\mathbf{k}_\parallel$ at the Fermi level E_F the transmission is given by

$$T\left(\mathbf{k}_\parallel, E_F\right) = \exp\left[-2\kappa\left(\mathbf{k}_\parallel\right)d\right]\left(\frac{4\kappa\left(\mathbf{k}_\parallel\right)k_\perp}{\kappa\left(\mathbf{k}_\parallel\right)^2 + k_\perp^2}\right)^2 ,$$

$$\kappa\left(\mathbf{k}_\parallel\right) = \sqrt{V - k_\perp\left(\mathbf{k}_\parallel\right)} \approx \kappa(0) + \frac{\mathbf{k}_\parallel^2}{2\kappa(0)} \quad k_\perp\left(\mathbf{k}_\parallel\right) = \sqrt{E_F - \mathbf{k}_\parallel^2} .$$

$$(3.20)$$

That means, with increasing in-plane wave vector the transmission decreases very rapidly due to the additional exponential factor $\exp\left(-\mathbf{k}_{\parallel}^2 d/2\kappa(0)\right)$. In this model, states with normal incidence contribute mostly to the TMR for large barrier thicknesses. This was found for the spin majority channel in ideal Fe/MgO/Fe junctions as shown in Fig. 3.17.

Analyzing the band structure of insulators, and especially the states with complex wave vector in the band gap, it was found that in all considered materials the states with the shortest decay rate $\kappa(0)$ are found in the center of the Brillouin zone [105]. So the matching of the states of electrode and barrier material in this high symmetric point is crucial for the tunnel current. This argument is much more restrictive at larger barrier thicknesses. The character and dispersion of the states in Fe and MgO in the center of the Brillouin zone are shown in Fig. 3.16. The states in the band gap of MgO with shortest complex wave vector have Δ_1 symmetry and match with the Δ_1 band in the majority band of Fe. An additional band with Δ_5 symmetry exists in MgO which fits to the corresponding band in the minority channel of Fe. In Ref. [90] the eigenstates of a Fe/MgO/Fe junction are computed and from the behavior of the probability amplitudes inside the MgO barrier the differences in the decay rates are obvious.

To illustrate, how many states are available in the Fe electrode to tunnel through the barrier, the number of states at the Fermi level with a given

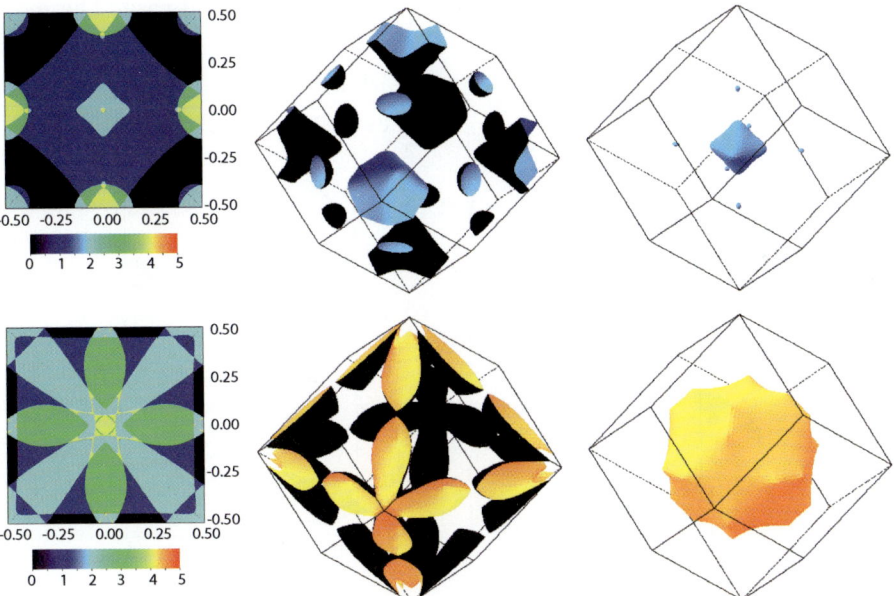

Fig. 3.18. Number of channels in the Fe(001) electrodes at the Fermi level in dependence on the in-plane wave vector $\mathbf{k}_{\parallel}$ for the minority (*top*) and the majority (*bottom*) electrons, the Fermi surfaces are given for illustration, from [26]

in-plane wave vector are shown in Fig. 3.18. The eight-fold symmetry of the Fe/MgO interface is evident. The number of states for an in-plane wave vector is given by the number of Fermi surface sheets, which are shown for illustration in the right panels. This can be seen from (3.14), when ballistic transport with in-plane momentum conservation is considered and all transmission coefficients are set to one. Then (3.14) counts the states traveling towards the barrier [116].

Including the electronic structure of the Fe leads and the MgO barrier the transmissions obtained by KKR methods confirm some aspects of the simple models, but also pronounced differences emerge. Figure 3.17 comprises two idependent calculations of Butler et al. [90] and of the author for Fe/MgO/Fe tunnel junction with a barrier thickness of 4ML. In the majority band the dominance of the states at the Brillouin center is confirmed. This is connected to the Δ_1 band shown in Fig. 3.16, middle panel. In contrast, the dominating minority states are located close to the Brillouin zone boundary, where a large density of states is found in the Fe interface atomic layer. Here, the interplay of the eigenstates of the leads, of the barrier and the structure of the interface becomes decisive for the appearance of highly transmissive channels.

To single out the features of the current-voltage characteristics we discuss in Fig. 3.19 the conductance g defined as the ratio of current and voltage $g = I/U$ and the TMR ratio as a function of external bias and the structure of the interfaces. Assuming perfect interfaces between Fe and MgO without any intermixed layers, this structure will be called the ideal case. The more

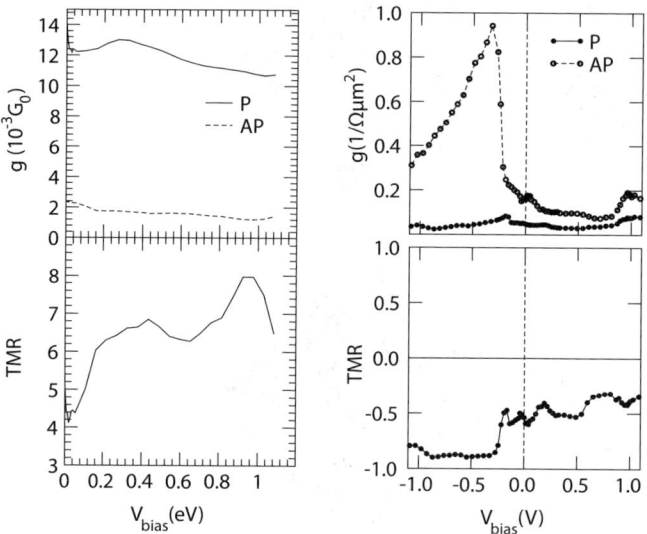

Fig. 3.19. Influence of interface structure on bias dependence of conductance and TMR in Fe/4MgO/Fe: left ideal interfaces; right one ideal and one decorated interface [119]

probable case of the creation of a mixed FeO layer at the bottom interface, as it was found in experiments [114], is called the asymmetric case, because for the top interface a perfect structure will be assumed. For the ideal case we obtain only small variations of the conductance with the voltage for both magnetic configurations. So, the resulting TMR is positive for all investigated bias values. Conductance for the P magnetic configuration is nearly constant in the asymmetric case. The strong reduction of the majority conductance by this mixed FeO layer was shown in calculations in Ref. [117]. The broken symmetry of the junction is reflected in the asymmetry for positive and negative voltages. In the AP configuration the conductance increases strongly for negative bias, that means that the current is injected from the electrode with the perfect interface. For all bias values a negative TMR ratio is obtained. This is in line with results from literature for larger barrier thicknesses [113].

It is evident, that the total conductance of a tunnel junction is equally determined by the electronic structure of leads and barrier, and the structure of the interfaces. The modification of the interfaces allows a generation of positive and negative TMR ratios. The bias characteristics is strongly influenced by the interface geometry, especially in the case of coherent transport, when the matching of the eigenstates mainly determines the tunnel probability. Enhancing the interface quality in experimental samples these theoretical predictions should be proven.

Ab initio electronic calculations allow for an important insight into the spin-dependent transport processes in magnetic tunnel junctions. It was demonstrated that Bloch electrons behave qualitatively different from free electrons. The wave functions inside the barrier and the matching conditions at the interfaces are quite different from those modeled by a rectangular barrier potential.

Applications

In principle, application areas for the TMR are the same as for the GMR effect, but at some fields the TMR elements can meet much stronger technological needs. This concerns the scalability of the effect with a shrinking element area and the temperature stability. The resistance of the elements increases with decreasing size and allows an easier embedding in the established CMOS technology.

For upcoming magnetic random access memory (MRAM) applications it is highly desirable to provide a large output voltage with small applied current. A single tunnel junction stores an information bit by the relative orientation of the electrode magnetizations. To read the information the resistance of the junction is evaluated. For the write process additional current lines have to be attached to provide the necessary magnetic field [118]. The prominent advantages in comparison to the established SRAM and DRAM technologies are the non-volatility, the resulting low power consumption, and low voltage. With respect to non-volatile semiconductor memory technologies a very fast access and a high endurance has to be emphasized. It is expected that the first commercial MRAM memory chips will be sold in 2005.

Recently nonvolatile MRAM devices are made from an array of magnetic tunnel junctions for special applications, but a broad application in mobile information technologies is envisaged.

References

1. European Commission, *Technology Road Map for Nanoelectronics* (2000). http://europa.eu.int/comm/research/briefings/nanotechnology_en.html
2. S.A. Wolf, D.D. Awschalom, R.A. Buhrman, J.M. Daughton, S. von Molnár, M.L. Roukes, A.Y. Chtchelkanova, D.M. Treger, Science 294, 1488 (2001)
3. J.F. Gregg, I. Petej, E. Jouguelet, C. Dennis, J. Phys. D 35, R121-155 (2002)
4. I. Žutic, J. Fabian, S. Das Sarma, Rev. Mod. Phys. 76, 323 (2004)
5. P. Grünberg, R. Schreiber, Y. Pang, M.B. Brodsky, H. Sowers, Phys. Rev. Lett. 57, 2442 (1986)
6. S.S.P. Parkin, N. More, K.P. Roche, Phys. Rev. Lett. 64, 2304 (1990)
7. P. Bruno, Phys. Rev. B 52, 411 (1995)
8. M.N. Baibich, J.M. Broto, A. Fert, F. Nguyen Van Dau, F. Petroff, P. Etienne, G. Creuzet, A. Friederich, J. Chazelas, Phys. Rev. Lett. 61, 2472 (1988)
9. G. Binasch, P. Grünberg, F. Saurenbach, W. Zinn, Phys. Rev. B 39, 4828 (1989)
10. A. Barthélémy, A. Fert, F. Petroff, in *Handbook of Magnetic Materials*, Vol.12, edited by K.H.J. Buschow, Elsevier, Amsterdam (1999)
11. E. Tsymbal, D.G. Pettifor, Solid State Phys. 56, 113 (2001)
12. E.Y. Tsymbal, O.N. Mryasov, P.R. LeClair, J. Phys. Cond. Matter 15, R109-142 (2003)
13. S. Datta, B. Das, Appl. Phys. Lett. 56, 665 (1990)
14. D.J. Monsma, J.C. Lodder, Th.J.A. Popma, B. Dieny, Phys. Rev. Lett. 74, 5260 (1995)
15. E.C. Stoner, Proc. R. Soc. London, Ser. A 165, 372 (1938)
16. O. Eriksson, A.M. Boring, R.C. Albers, G.W. Fernando, B.R. Cooper, Phys. Rev. B 45, 2868 (1992)
17. H. Hasegawa, J. Phys. F 16, 1555 (1986)
18. V. Bellini, N. Papanikolaou, R. Zeller, P.H. Dederichs, Phys. Rev. B 64, 094403 (2001)
19. J. Korringa, Physica 13, 392 (1947)
20. W. Kohn, N. Rostoker, Phys. Rev. 94, 1111 (1954)
21. Lord Rayleigh, Phil. Mag. 34, 481 (1892)
22. P.P. Ewald, Annals of Physics 49, 1 (1916); do. 49, 117 (1916)
23. W.H. Butler, P.H. Dederichs, A. Gonis, R.L. Weaver *Application of Multiple Scattering Theory to Material Science*, MRS Symposia Proceedings No 253, Materials Research Society, Pittsburgh (1992)
24. N. Papanikolaou, R. Zeller, P.H. Dederichs, J. Phys. Cond. Matter 14, 2799 (2002)
25. W. Kohn, L.J. Sham, Phys. Rev. 140, A1133 (1965)
26. http://www.physik.tu-dresden.de/~fermisur (2006)

27. J. Unguris, R.J. Celotta, D.T. Pierce, Phys. Rev. Lett. 67, 140 (1991)
28. J. Unguris, R.J. Celotta, D.T. Pierce, Phys. Rev. Lett. 69, 1125 (1992)
29. S.S.P. Parkin, Phys. Rev. Lett. 67, 3598 (1991)
30. M.A. Ruderman, C. Kittel, Phys. Rev. 96, 99 (1954), T. Kasuya, Progr. Theor. Phys. (Japan) 16, 45 (1956), K. Yosida, Phys. Rev. 106, 893 (1957)
31. J. Friedel, Phil. Mag. 43, 153 (1952)
32. P. Bruno, C. Chappert, Phys. Rev. Lett. 67, 1602, 2592 (E) (1991)
33. P. Bruno, C. Chappert, Phys. Rev. B 46, 261 (1992)
34. A. Fuš, S. Demokritov, P. Grünberg, W. Zinn, J. Magn. Magn. Mater. 103, L221 (1992)
35. M.T. Johnson, S.T. Purcell, N.W.E. McGee, R. Coehoorn, J. aan de Stegge, W. Hoving, Phys. Rev. Lett. 68, 2688 (1992)
36. J. Unguris, R.J. Celotta, D.T. Pierce, J. Appl. Phys. 75, 6437 (1994)
37. J. Opitz, P. Zahn, J. Binder, I. Mertig, Phys. Rev. B 63, 094418 (2001)
38. J. Unguris, R.J. Celotta, D.T. Pierce, Phys. Rev. Lett. 79, 2734 (1997)
39. F. Herman, J. Sticht, M. v. Schilfgaarde, J. Appl. Phys. 69, 4783 (1991)
40. M.D. Stiles, Phys. Rev. B 48, 7238 (1993)
41. P. Lang, L. Nordström, R. Zeller, P.H. Dederichs, Phys. Rev. Lett. 71, 1927 (1993)
42. J. Kudrnovský, V. Drchal, I. Turek, P. Weinberger, Phys. Rev. B 50, 16105 (1994)
43. A. Fert, P. Bruno, in B. Heinrich and J.A.C. Bland (eds.), *Ultrathin Magnetic Structures*, Springer, Berlin, (1994)
44. R. Kubo, J. Phys. Soc. Japan 12, 570 (1957)
45. I. Mertig, Rep. Prog. Phys. 62, 237 (1999)
46. P. Zahn, I. Mertig, M. Richter, H. Eschrig, Phys. Rev. Lett. 75, 2996 (1995)
47. W.H. Butler, X.-G. Zhang, D.M.C. Nicholson, J.M. MacLaren, Phys. Rev. B 52, 13399 (1995)
48. J. Mathon, Phys. Rev. B 55, 960 (1997)
49. P. Weinberger, Phys. Rev. 377, 281-387 (2003)
50. A.I. Yanson, G. Rubio-Bollinger, H.E. van den Brom, N. Agraït, J.M. van Ruitenbeek, Nature 395, 783 (1998)
51. H. Ohnishi, Y. Kondo, K. Takayanagi, Nature 395, 780 (1998)
52. S.S.P. Parkin, C. Kaiser, A. Panchula, P.M. Rice, B. Hughes, M. Samant, S.-H. Yang, Nature Materials 3, 862 (2004)
53. S. Yuasa, T. Nagahama, A. Fukushima, Y. Suzuki, K. Ando, Nature Materials 3, 868 (2004)
54. L. Piraux, J.M. George, J.F. Despres, C. Leroy, E. Ferain, R. Legras, K. Ounadjela, A. Fert, Appl. Phys. Lett. 65, 2484 (1994)
55. R.H.M. Smit, Y. Noat, C. Untiedt, N.D. Lang, M.C. Van Hemert, J.M. Van Ruitenbeek, Nature 419, 906 (2002)
56. R. Landauer, Phil. Mag. 21, 863 (1970)
57. M. Büttiker, Y. Imry, R. Landauer, S. Pinhas, Phys. Rev. B 31, 6207 (1985)
58. J.M. MacLaren, X.G. Zhang, W.H. Butler, X. Wang, Phys. Rev. B 59, 5470 (1999)
59. E.Yu. Tsymbal, D.G. Pettifor, J. Appl. Phys. 85, 5801 (1999)
60. D. Wortmann, H. Ishida, S. Blügel, Phys. Rev. B 65, 165103 (2002)
61. I. Mertig, R. Zeller, P.H. Dederichs, Phys. Rev. B 47, 16178 (1993)
62. I. Mertig, P. Zahn, M. Richter, H. Eschrig, R. Zeller, P.H. Dederichs, J. Magn. Magn. Mater. 151, 363 (1995)

63. P. Zahn, J. Binder, I. Mertig, R. Zeller, P.H. Dederichs, Phys. Rev. Lett. 80, 4309 (1998)
64. N.C. Mott, Adv. Phys. **13**, 325 (1964)
65. R. Landauer, IBM J. Res. Dev. 32, 306 (1988)
66. M. Büttiker, IBM J. Res. Dev. 32, 317 (1988)
67. S. Datta, *Electronic Transport in Mesoskopic Systems*, Cambridge University Press, Cambridge (1995)
68. D.S. Fisher, P.A. Lee, Phys. Rev. B 23, 6851 (1981)
69. H.U. Baranger, A.D. Stone, Phys. Rev. B 40, 8169 (1989)
70. Ph. Mavropoulos, N. Papanikolaou, P.H. Dederichs, Phys. Rev. B 69, 125104 (2004)
71. Databook Semiconductor Sensors, Infineon Technologies AG, Munich (2000)
72. T. Oguchi, J. Magn. Magn. Mater. 126, 519 (1993)
73. R.E. Camley, J. Barnaś, Phys. Rev. Lett. 63, 664 (1989)
74. P.M. Levy, S. Zhang, A. Fert, Phys. Rev. Lett. 65, 1643 (1990)
75. T. Valet, A. Fert, Phys. Rev. B 48, 7099 (1993)
76. S.S.P. Parkin, Phys. Rev. Lett. 71, 1641 (1993)
77. C.H. Marrows, B.J. Hickey, Phys. Rev. B 63, 220405 (2001)
78. Z.Q. Qiu, J. Pearson, S.D. Bader, Phys. Rev. B 46, 8659 (1992)
79. P.J.H. Bloemen, M.T. Johnson, M.T.H. van de Vorst, R. Coehoorn, J.J. de Vries, R. Jungblut, J. aan de Stegge, A. Reinders, W.J.M. de Jonge, Phys. Rev. Lett. 72, 764 (1994)
80. P. Lang, PhD Thesis, RWTH Aachen (1995)
81. F. Giron, P. Boher, P. Houdy, P. Beauvillian, K. Le Dang, P. Veillet, J. Magn. Magn. Mater. 121, 24 (1993)
82. P.H. Dederichs, R. Zeller, in *Festkörperprobleme (Advances in Solid States Physics)*, Volume XXI, pp. 243-269, Vieweg, Braunschweig (1981)
83. J. Binder, P. Zahn, I. Mertig, J. Appl. Phys. 89, 7107 (2001)
84. W.H. Butler, X.-G. Zhang, D.M.C. Nicholson, T.C. Schulthess, J.M. MacLaren, Phys. Rev. Lett. 76, 3216 (1996)
85. P. Zahn, J. Binder, I. Mertig, Phys. Rev. B 68, 100403(R) (2003)
86. D.J. Larson, A.K. Petford-Long, A. Cerezo, G.D.W. Smith, D.T. Foord, T.C. Anthony, Appl. Phys. Lett. 73, 1125 (1998)
87. J. Schleiwies, G. Schmitz, S. Heitmann, A. Hütten, Appl. Phys. Lett. 78, 3439 (2001)
88. http://ssdweb01.storage.ibm.com/hardsoft/diskdrdl/technolo/gmr/gmr.htm
89. J.S. Moodera, L.R. Kinder, T.M. Wong, R. Meservey, Phys. Rev. Lett. 74, 3273 (1995)
90. W.H. Butler, X.-G. Zhang, T.C. Schulthess, J.M. MacLaren, Phys. Rev. B 63, 054416 (2001)
91. T. Miyazaki and N. Tezuka, J. Magn. Magn. Mater. 139, L231 (1995)
92. P.M. Tedrow, R. Meservey, Phys. Rev. Lett. 26, 192 (1971)
93. P.M. Tedrow, R. Meservey, Phys. Rev. B 7, 318 (1973)
94. M. Julliere, Phys. Lett. 54A, 225 (1975)
95. R. Meservey, P.M. Tedrow, Physics Reports 238, 173-243 (1994)
96. J. Faure-Vincent, C. Tiusan, E. Jouguelet, F. Canet, M. Sajieddine, C. Bellouard, E. Popova, M. Hehn, F. Montaigne, A. Schuhl, Appl. Phys. Lett. 82, 4507 (2003)
97. J.M. De Teresa, A. Barthélémy, A. Fert, J.P. Contour, R. Lyonnet, F. Montaigne, P. Seneor, A. Vaurès, Phys. Rev. Lett. 82, 4288 (1999)

98. P. LeClair, H.J.M. Swagten, J.T. Kohlhepp, R.J.M. van de Veerdonk, W.J.M. de Jonge, Phys. Rev. Lett. 84, 2933 (2000)

99. P. LeClair, J.T. Kohlhepp, H.J.M. Swagten, W.J.M. de Jonge, Phys. Rev. Lett. 86, 1066 (2001)

100. J.C. Slonczewski, Phys. Rev. B 39, 6995 (1989)

101. S. Zhang, P.M. Levy, Phys. Rev. Lett. 81, 5660 (1998)

102. O. Wunnicke, N. Papanikolaou, R. Zeller, P.H. Dederichs, V. Drchal, J. Kudrnovsk'y, Phys. Rev. B 65, 064425 (2002)

103. A.D. Stone, P.A. Lee, Phys. Rev. Lett. 54, 1196 (1985)

104. V. Heine, Proc. Phys. Soc. 81, 300 (1963)

105. Ph. Mavropoulos, N. Papanikolaou, P.H. Dederichs, Phys. Rev. Lett. 85, 1088 (2000)

106. J.M. MacLaren, X.-G. Zhang, W.H. Butler, Phys. Rev. B 56, 11827 (1997)

107. J. Mathon, A. Umerski, Phys. Rev. B 63, 220403(R) (2001)

108. S. Zhang, P.M. Levy, A.C. Marley, S.S.P. Parkin, Phys. Rev. Lett. 79, 3744 (1997)

109. M. Sharma, S.X. Wang, J.H. Nickel, Phys. Rev. Lett. 82, 616 (1999)

110. C. Tiusan, J. Faure-Vincent, C. Bellouard, M. Hehn, E. Jouguelet, A. Schuhl, Phys. Rev. Lett. 93, 106602 (2004)

111. H.F. Ding, W. Wulfhekel, J. Henk, P. Bruno, J. Kirschner, Phys. Rev. Lett. 90, 116603 (2003)

112. X.-G. Zhang, W.H. Butler, arXiv:cond-mat/0409155 (2004)

113. C. Zhang, X.-G. Zhang, P.S. Krstić, Hai-ping Cheng, W.H. Butler, J.M. MacLaren, Phys. Rev. B 69, 134406 (2004)

114. H.L. Meyerheim, R. Popescu, J. Kirschner, N. Jedrecy, M. Sauvage-Simkin, B. Heinrich, R. Pinchaux, Phys. Rev. Lett. 87, 076102 (2001)

115. H.L. Meyerheim, R. Popescu, N. Jedrecy, M. Vedpathak, M. Sauvage-Simkin, R. Pinchaux, B. Heinrich, J. Kirschner, Phys. Rev. B 65, 144433 (2002)

116. K.M. Schep, P.J. Kelly, G.E.W. Bauer, Phys. Rev. Lett. 74, 586 (1995)

117. X.-G. Zhang, W.H. Butler, J. Phys. Cond. Matter 15, R1603 (2003)

118. http://domino.research.ibm.com/comm/pr.nsf/pages/rsc.mramimages.html (2005)

119. C. Heiliger, P. Zahn, B.Yu. Yavorsky, I. Mertig, Phys. Rev. B 72, 180406(R) (2005)

4

Theoretical Investigation of Interfaces

Sibylle Gemming and Michael Schreiber

Summary. The proper treatment of defects is one of the major tasks in materials design, because defects are responsible for the either desirable or detrimental deviations between the characteristics of the material to be tuned and the well-known properties of an ideal crystal. Microelectronic devices work because of clever point defect engineering, line defects govern plastic deformation processes, and interfaces determine the mechanical stability of composite materials. Especially interfaces gain importance with the current trend towards nanoscale materials; first, the surface-to-volume ratio is strongly increased in nanocrystalline material, and, second, stable arrangements of point or line defects require a minimum crystallite size, which can be larger than the actual nanocrystallites. Thus, the present chapter gives an introduction into the most common approaches for modeling interface properties. We introduce the basic concepts of interface symmetry, structure and analysis with a strong focus on the theoretical methods and give an overview of currently available techniques for the modeling and simulation of the interface properties at an atomic-scale level. Two fundamentally different interface types are distinguished: The discussion of the homophase boundary properties is focussed on oxide grain boundaries, which we studied extensively in comparison with amply available experimental observations. For the heterophase boundaries examples of non-reactive, reactively doped, and inherently reactive boundaries are presented. A special focus lies on the interfaces between metals and oxides where the discrepancy of the material properties across the interface is most prominent and all three bonding situations can occur: weak adhesion between inert fragments, activated adhesion upon doping, and strong adhesion.

4.1 Interfaces – Boundaries Between Two Phases

Intrinsic or specially designed interfaces are present in most of the objects which we handle in every-day life. Any vehicle we use for transportation provides examples for interface engineering with the aim of improving the properties of the raw material. For instance, proper doping of grain boundaries in steel with additional transition elements makes the frame of a bicycle more stiff or less prone to corrosion, or the proper combination of platinum metal and ceramic carrier substrate may make the catalyst of a car more long-lived. Thus, it is of fundamental importance to gain a thorough understanding of the structure-property relationship at internal boundaries. For a start we focus on the structure part of this demanding task and briefly become acquainted with the terminology developed to characterise interface structures. Then the major interaction mechanisms at an interface are discussed. For a more detailed view of the subject the reader is referred to the textbook of Sutton and Balluffi [1].

4.1.1 Introduction

A basic distinction is made between homophase and heterophase boundaries. *Homophase boundaries* are interfaces between two crystallites of the same material with the same crystal structure and the same elemental composition. Since this type of boundary was studied most extensively in polycrystalline materials, the term *grain boundary* is used synonymously to homophase boundary. An example for this boundary type is the $\Sigma 3(111)[\text{-}110]$ twin grain boundary in $SrTiO_3$ described in Sect. 4.3 [2, 3]. All other cases belong to the group of *heterophase boundaries* or shortly and less precise of general *interfaces*. This group comprises not only boundaries between two different materials, as for instance in the metal–ceramic boundaries of the adhesion system Al or Ag on spinel [4–9]. In the strict sense, also boundaries between two different modifications with the same composition, such as the boundary line of the structural phase transition from γ- to α-Al_2O_3 [10] are heterophase boundaries.

Next, the question shall be tackled, which geometrical degrees of freedom exist at a phase boundary, because they provide a starting point for the optimisation of interface properties. First, the two crystallites which form the boundary have a certain relative orientation in space. This provides three degrees of freedom: two for the specification of a joint rotation axis and one for the rotation angle. When the crystals are properly oriented, two further degrees of freedom are given by the possibility to choose two directions, which determine the orientation of the boundary plane. Figure 4.1 gives a schematic representation.

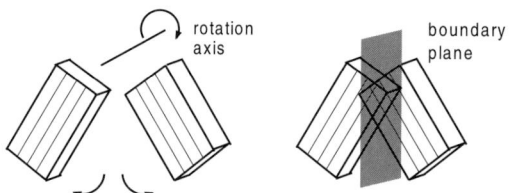

Fig. 4.1. The five macroscopic degrees of freedom at a boundary between two crystallites

These five degrees of freedom do not depend on the specific crystal structure of the material, thus they are termed *macroscopic* degrees of freedom.

Additional *microscopic* degrees of freedom are introduced by the particulars of the crystal structure. The atoms in the two bicrystal halves can be shifted with respect to each other by a rigid-body translation. This introduces the translation state T which is characterised by the translation vector t. t is independent of the local atom rearrangements in the close vicinity of the boundary, but denotes the relative shift of the bulk lattices of the two adjoining phases. The basis vectors for t are usually chosen in such a manner that two of its components are parallel to the interface and one is perpendicular. The translation vector hence contains three microscopic degrees of freedom, as shown schematically in Fig. 4.2.

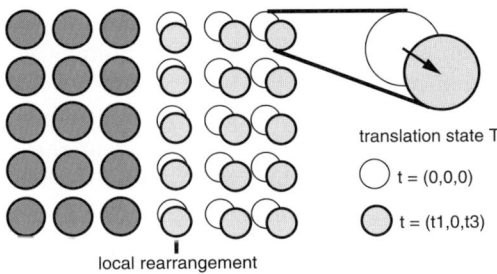

translation state T

$t = (0,0,0)$

$t = (t1,0,t3)$

local rearrangement

Fig. 4.2. Translation state and translation vector. Note that atom displacements at the boundary do not influence the translation state. The region of the top right atom is enlarged to indicate the translation vector

Two further microscopic degrees arise, when the crystallites contain more than just one crystallographically equivalent site. In those cases, the termination planes of the two crystallites, which form the interface, can exhibit several different compositions, thus, the interface stoichiometry provides an additional degree of freedom. In part, the choice of the termination planes can be implicitly treated by a proper component of the translation vector perpendicular to the interface, since this corresponds to shifting the basis in one of the two crystallites. Thus, shifting the basis on the other side leads to this additional degree of freedom. This situation is schematically represented in Fig. 4.3. Examples are given in Refs. [3, 11], where two different termination planes had to be distinguished. Last, in chiral crystals, the possibility to form four enantiomorphic combinations of right- and left-handed crystallites yields another degree of freedom, such that altogether up to 5 microscopic degrees of freedom can occur.

Especially at interfaces between two different phases the lattice constants of the adjoining lattices do not necessarily match. In the extreme case this leads to an aperiodic arrangement of the atoms adjacent to the boundary

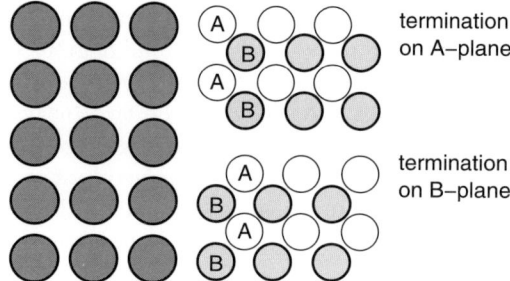

Fig. 4.3. Choice of the termination plane in the case of a crystal with a diatomic basis

plane, and the interface is termed *incommensurate*. If some long-range, one- or two-dimensional ordering is obtained at the boundary, one denotes it as a *commensurate* boundary. Finally, the most well-defined translation state is obtained for a *coherent* interface, where the continuity of a common reference lattice is maintained across the boundary throughout the whole bicrystal.

The structure of coherent boundaries can be derived with the help of the coincidence site lattice (CSL). The major steps of this procedure are sketched in Fig. 4.4. In order to generate the CSL, the two crystals, here in black and white, are rotated with respect to each other and superimposed such that white and black spots interpenetrate and form the so-called dichromatic pattern. Most of the black and white circles do not coincide. The (normally fewer) matching points are called the coincidence points, and they form the CSL with a lattice constant a_{CSL}. a_{CSL} is usually larger than the lattice constants of the two adjoining crystals. Thus, a measure for the density of coincidence points in the boundary plane is given by the Σ-*value*, a concept which was derived for cubic crystal systems. To obtain the Σ-value one calculates the area of the CSL repeat unit and divides it by the smaller area of the unit cell of the constituent crystals.

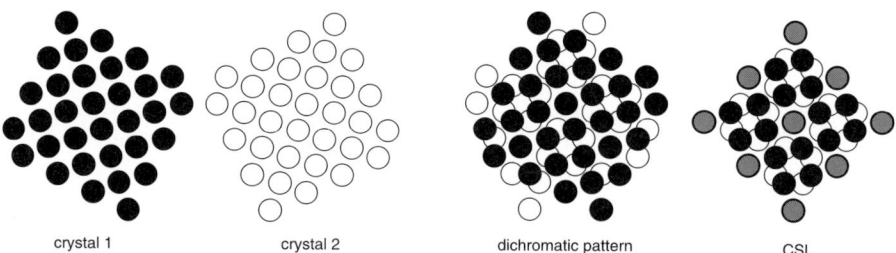

Fig. 4.4. Construction of the dichromatic pattern and the coincidence site lattice. The CSL plot shows a slightly different part of the two original crystals. The coincidence points are drawn in grey

If the lattice constants of the adjoining crystals match quite closely, then local strains at the interface can lead to an elastic deformation at the interface, which establishes the perfect CSL arrangement. Such a situation is called a near-CSL boundary, and the deviation from the perfect lattice coincidence is given by the *lattice mismatch* or *misfit*

$$\delta = \frac{|a_0^1 - a_0^2|}{a_0^1 + a_0^2} \,,$$

where a_0^1 and a_0^2 are the lattice constants of the two crystals at the boundary.

4.1.2 Interactions

Theoretical investigations of interfaces are carried out to elucidate the major interaction types, which determine the atomistic structure at an interface, and the relation of these interaction mechanisms to bulk properties. In this way, a qualitative prediction of the interface behaviour has become possible on the basis of data obtained for the unperturbed bulk. This approach has the advantage that one does not have to model or measure the precise details of a specific boundary in order to understand or predict its geometry, stability, reactivity, or electronic structure. Rather a simplified model may be built, which yields trends on the basis of precise bulk data.

Five major interaction terms have been obtained from the investigations of various homophase and heterophase boundaries [2–9, 11–13] and correlated to bulk properties:

(1) Coulomb interaction between the (partial) charges in ionic crystals
(2) elastic interactions due to misfit strains at the interface
(3) electron transfer processes across the interface, when the atoms at the interface exhibit different electronegativity
(4) Pauli repulsion between filled valence shells
(5) image charge interactions between fixed (ionic) charges and induced dipoles at the other side of the boundary

The first two interactions determine the structure of pure, stoichiometric homophase boundaries, the other three factors come into play, when elemental excess of one atom type, additional dopants, or a second phase is present at the boundary. These five interaction terms shall be discussed in more detail.

Coulomb Interaction

Especially in ionic materials, such as MgO or Al_2O_3 with high formal charges of +3 (Al), +2 (Mg), and −2 (O), the Coulomb interaction is the dominant structure-determining factor. The calculated charges are usually lower than the formal ones and depend on the specific procedure how the electron density is spatially decomposed. As the partial charge of a single atom within an extended solid is not an experimentally observable quantity, there is no unique ansatz for the derivation of partial charges, but rather a large variety of methods which range from a simple spatial decomposition to methods which are based on physical plausibility arguments. To mention some of the most common approaches: the site-specific projection onto numerical basis functions clearly just yields a spatial decomposition with a high degree of arbitrariness, and the values obtained can be compared safely only within a set of calculations that employs the same numerical basis. The projection onto spherical harmonics or onto localised atomic orbitals, e. g. in the crystal orbital overlap population (COOP) [14], provides the possibility to choose the radius of the projection sphere such that it follows the minimum of the electron density. These procedures are applicable to spherical ions like M^{2+} in the perovskites $MTiO_3$, M = Sr, Ba, or $MgAl_2O_4$ [2, 14], but must be modified to account for deviations from the spherical symmetry. The modification can be achieved by decomposition of the electron density in space-filling Voronoi polyhedra following the minimum of the density gradient by a Bader-type topological analysis [15]. A completely different approach is the derivation of Born effective charges, which are obtained as reaction of the electronic system to an external perturbation such as a static electric field [16]. Despite this variety of approaches a simple theory based on the Coulomb interaction between the formal charges can be employed for a quick screening of different structure models and yields potentially low-energy grain boundaries [17].

Elastic Interaction

In contrast to the Coulomb interaction of ions, which can be both attractive or repulsive, the elastic contributions to the interface interaction reduce the adhesion, because at least one side of the interface cannot maintain its optimum lattice parameter, but is under compressive or tensile stress. Both cases lead to deviations from the minimum of the equation of state for the bulk crystal, and the cohesive energy in the boundary region is diminished. At grain boundaries these stresses can be released by the formation of misfit dislocations close to the boundary [18, 19], or also by the presence of glassy films at the boundary, especially when dopants are added (e. g. Mg in polycrystalline Al_2O_3 [20]). In adhesive thin-film heterophase systems the film can even undergo a misfit-induced dewetting transition at higher film thickness, when the deformation energy exceeds the bonding energy across the interface. An illustrative example is the growth of Ge on Si, where only the first Ge layer can bond epitaxially to the Si substrate, and then the growth mode changes to island formation. Due to the covalent Si–Ge interaction the dewetting transition is not complete, but rather leads to this Stranski–Krastanov-type behaviour [21]. The bulk quantities to estimate the elastic contribution are the lattice constants

a, the misfit δ, and the bulk modulus B or the stress tensor σ. Since these quantities are properties, which depend on the presence of more than one atom, a further reduction to a representative atomistic quantity is not viable.

Electron Transfer

Especially at metal-nonmetal heterophase boundaries two electronically very dissimilar materials meet each other: the metal valence and conduction band states are delocalised and no band gap separates them. In the semiconductor or ionic solid, an energy gap separates occupied valence and empty conduction states, and especially in the ionic solid the electrons are more strongly localised and the corresponding bands are flat. When the two materials are brought into contact, the Fermi levels of both sides align. In a rigid band model this alignment can lead to an electron transfer across the boundary (see panel (a) of Fig. 4.5) and to the generation of localised, metal-induced states in the band gap [22] as depicted in panel (b). In this way, strong, directed, chemical metal-to-oxygen bonding was monitored, for instance, for interfaces between a main-group metal (Al) or an early transition metal (Nb, Ti) and a main group oxide such as MgO or Al_2O_3 [23–26]. Another example is the silicide formation at reactive boundaries between a Si(111) or Si(110) surface and light transition or rare-earth elements [27–29]. The corresponding bulk quantity is the electronegativity, which specifies the ability of an atom to retain electrons in its uppermost occupied states. In a quantum chemical treatment the electronegativity is obtained as average of ionisation energy and electron affinity, i. e. one half of the difference between the total energies of cation and anion [30]. Although the electronegativity of an atom changes in its chemical environment, the comparison of the atomic electronegativities allows an estimate about the extent and the direction of the electron transfer.

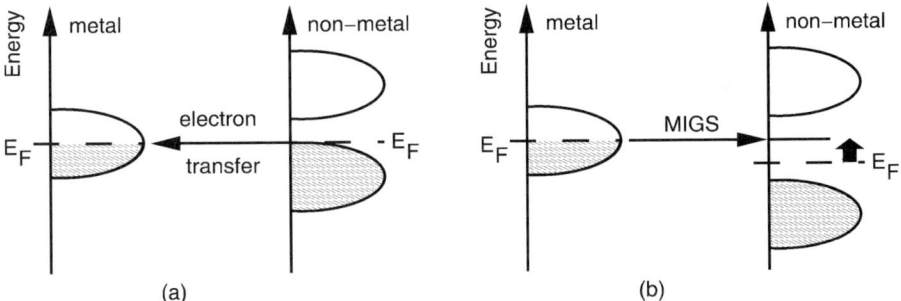

Fig. 4.5. Band alignment at a boundary between a metal and a non-metal for (**a**) electron transfer and (**b**) metal-induced gap states (MIGS)

Pauli Repulsion

In the band picture discussed above it is immediately obvious that the electron transfer is not favourable when both materials have filled and flat bands. In this case, no electron transfer can yield an additional stabilisation. Quite on the contrary the closed shells experience a Pauli repulsion when brought in close contact, and the adhesion is diminished. An atomistic quantity which yields the propensity of a system for Pauli repulsion is the electron gas parameter r_s, which can be calculated to high accuracy by a precise numerical solution of the Schrödinger or Dirac equation for the single atom. The electron gas parameter specifies the effective radius, which contains one electron of the given atom, thus it is inversely proportional to the average electron density. Small values of r_s characterise the atoms with a high electron density, which are most prone to experience Pauli repulsion.

Image Charge Interaction

If the adhesion of a metal with an ionic solid is more strongly influenced by the Pauli repulsion (4) than by the electron transfer bonding (3) there may be another weakly bonding interaction. This bonding is caused by the attractive interaction between the fixed charges of the ionic solid and the induced multipoles in the metal, the so-called *image charges*. Hence, this interaction is always (weakly) attractive. In conjunction with the interactions (1) to (4) this requires that the electronic bands at the metal site are almost completely filled but not too strongly localised, and can redistribute upon interaction with the charges of the ionic crystal. The induction of electrostatic image charges in the metal by the vicinity of an ionic crystal has been discussed widely as an alternative source of attractive Coulomb interactions in weakly bound systems [31–35]. As with the elastic interactions in (2), the definition of a specific atomistic quantity is difficult, since induced dipoles are per se a quantity which spans over more than one atom. One may, however, choose as key quantity the dispersion of the valence/conduction states, which can be obtained from a calculation or measurement of the unperturbed bulk metal. A high dispersion is characteristic of nearly free conduction electrons, e. g. the $5s^1$ and $6s^1$ states in Ag and Au, whereas a low dispersion characterises more localised states, like the d levels. Thus, a high band dispersion indicates that the metal electrons are mobile enough to rearrange in reaction to an external array of fixed charges.

4.2 Theoretical Methods

This section provides an overview over the most common approaches to model the structural and electronic properties of interfaces at the microscopic level.

Some aspects of density-functional (DF) band-structure theory are briefly reviewed, because density-functional theory (DFT) has proven an extremely efficient approach to obtain the properties of a given model system in its electronic ground state. For a more thorough introduction to DFT and its extensions to non-equilibrium systems the reader is referred to review articles such as Refs. [36–38] or textbooks such as Refs. [39, 40]. Second, an overview over classical, atomistic modelling will be given, which introduces the compressible ion model for ionic solids, the embedded atom approach, which is especially well suited for metals, and the image-charge model for heterophase boundaries between a metal and an ionic crystal (see also Ref. [1]). An exhaustive review of interface simulation should also mention the ample field of mesoscopic modelling with the help of continuum methods, e.g. described in a recent review by Stoneham and Harding [41]. For a more detailed description of modelling interatomic forces in materials the recent book by Finnis is recommended [42].

4.2.1 Theory of the Electronic Structure

Density-Functional Theory

Common quantum chemical approaches to the "ab-initio", i.e. the parameter-free treatment of electronic properties employ a wave function $|\Psi\rangle$ to describe all relevant electronic degrees of freedom. Thus, this wave function $|\Psi\rangle$ depends on the spatial variables of all interacting particles and computations involving the wave function soon become rather demanding as the system size increases. DFT provides an elegant means to circumvent this problem. As shown by Hohenberg and Kohn [43] in two theorems the knowledge of the ground-state electron density $\rho_0(\boldsymbol{r})$ is sufficient for the derivation of all observable ground-state properties. The first theorem states that the total energy of the electronic ground state $E_0^{\mathrm{el}}(\{\boldsymbol{R}_\alpha\})$ as well as other ground-state observables are functionals of the ground-state electron density $\rho_0(\boldsymbol{r})$ only, i.e. $E_0^{\mathrm{el}}(v_{\mathrm{ext}}\{\boldsymbol{R}_\alpha\}) = E^{\mathrm{el}}[\rho_0(\boldsymbol{r})]$.

Once the correct form of the functional and $\rho_0(\boldsymbol{r})$ are known it is straightforward to calculate the corresponding observable. For an exclusively computational approach, however, this first theorem becomes powerful only in conjunction with the second Hohenberg–Kohn theorem: It states that for a fixed external potential $v_{\mathrm{ext}}(\boldsymbol{r})$ the functional $E^{\mathrm{el}}[\rho(\boldsymbol{r})]$ exhibits its global minimum when $\rho(\boldsymbol{r})$ equals $\rho_0(\boldsymbol{r})$. Thus, the ground-state density can be obtained by minimising the functional $E^{\mathrm{el}}[\rho(\boldsymbol{r})]$ with the help of a variational scheme under the constraint that the number of electrons remains constant. In the original work these theorems were proven only under specific constraints, which were later alleviated by Levy [44]. Further generalisations were introduced for spin-polarised systems [45], for systems at finite temperature [46], and for relativistic systems [47, 48].

Computational Details of Bulk State Calculations

For an adequate treatment of an extended, crystalline solid one can make use of the translational symmetry. In this manner, only a small unit cell of the whole crystal has to be calculated explicitly, whereas the remainder can be constructed by applying periodic boundary conditions in three dimensions. This implies that the effective potential, the wave function, and thus also the electron density must obey certain symmetry requirements with respect to translations by any combination or multiple of the lattice vectors which span the unit cell. Several proofs of this theorem, called *Bloch's theorem*, exist and can be found in textbooks on solid state physics, e. g. in Ref. [50].

The requirement of lattice periodicity immediately suggests plane waves as an optimal basis for the representation of the electronic states. However, the strongly localised inner core electrons, such as $1s$ of Mg or Al, require a very high number of plane waves (or kinetic energy cutoff) for an adequate description of the short-range oscillations in the vicinity of the nucleus. This problem can be overcome by different strategies:

- by the use of localised, hydrogen-like functions within a sphere around the atomic site (augmented plane wave, APW methods),
- by the pseudopotential technique, which treats only the valence and semi-core states explicitly in a plane-wave basis, while an effective core potential accounts for the core-valence interaction (PW-PP methods, among them the Car-Parrinello approach [49]),
- or by the mixed-basis pseudo-potential (MB-PP) technique, which augments the plane-wave basis by non-overlapping atom-centered local functions that represent density oscillations close to the core.

A very successful compromise between efficiency and accuracy is provided by the use of norm-conserving pseudopotentials, which yield the same electron scattering as the all-electron potential by construction. With pseudopotentials also the relativistic contraction of inner shells in heavy atoms and its effect on the valence electrons can be included scalar-relativistically. In this way, also up to several hundred heavy atoms can be included in the model supercells for the analysis of interfaces.

Density-Functional Tight-Binding Approaches

Larger systems of up to several thousands of atoms are still tractable at an electronic structure level if simplifications of the full DFT scheme are introduced, which lead to a tight-binding (TB) description. In the standard TB method the electronic valence states are represented as a symmetry-adapted superposition of – usually rather few – orthogonalized atomic-like orbitals. The exact many-body Hamilton operator is approximated by a parametrised Hamiltonian matrix, whose matrix elements are fitted to the band structure of a suitable reference system. An additional short-range repulsive two-particle interaction includes the ionic repulsion and corrections due to approximations made in the band-structure term. It can be determined as a parametrised function of the interatomic distance, which reproduces the cohesive energy and elastic constants like the bulk modulus for crystalline systems. Thus, the results of a TB calculation may depend crucially on the parametrisation.

Nevertheless successful application has been made to high accuracy band structure evaluations [51], band calculations in semiconductor heterostructures [52], device simulations for optical properties [53], simulations of amorphous solids [54], and predictions of low-energy silicon clusters [55, 56] (for a review, see Ref. [57]). If a delicate charge balance is required to establish bonding between different atom types an adjustment of the charge distribution can be achieved via a self-consistent-field (SCF) treatment.

Due to the simple parametrisation, the TB methods allow routine calculations with up to 1000 to 2000 atoms. Thus they allow an extension to more irregular interface structures, which have a larger repeat unit. Another advantage is the simple derivation of additional quantities from TB data, because the TB approximations also simplify the mathematical effort for the calculation of material properties. E.g., the extension to the time-dependent DF response theory allowed the calculation of the optical and vibrational spectra of large organic molecules [58]. Even the transport in a nanodevice of conducting and semiconducting segments along a nanotube could be modelled [59].

4.2.2 Classical Modelling

Image-Charge Models

Image charges occur predominantly at metal–ceramic boundaries. As discussed above, image charges stem from a polarisation of the metal induced by the fixed positive and negative charges in the ionic solid. Figure 4.6 schematically shows this process for electron removal from a metal surface to a given position z_0 and the induced positive charge $(-Q)$ below the surface at $-z_0$.

For increasing distance above the surface $Z > z_0$ the potential decreases exponentially, thus the image charge is a rather short-ranged quantity. Parallel to the interface with increasing value of r, however, the decay is determined by the classical $1/r$ dependency for Coulomb-type interactions. Here, however, this decay is additionally modulated by the cosine function. These concepts have been generalised to treat the reaction of the metal electrons to an array of ionic charges [31, 34]. However, the position of the image plane at the metal side is uniquely defined only for the unperturbed jellium model [60], and experimental data must be employed to derive the position of z_0 in cubic metals [61].

It was furthermore derived, that the image-charge contribution equals one half of the electrostatic interaction energy between the induced electron density change and the potential of the external electron [34]. Thus, if no other

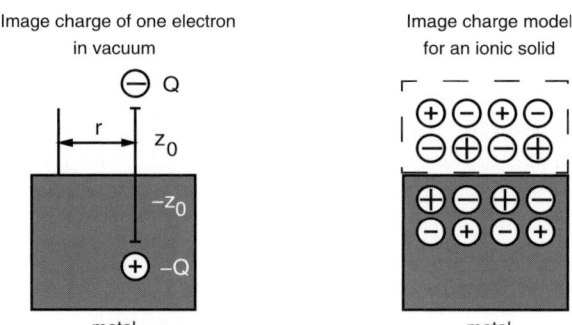

Fig. 4.6. Image charge formation at a metal–ceramic interface

bonding mechanisms are possible, the image charge interaction is the dominant, attractive interaction. Image charge models have, therefore, been applied especially to boundaries between noble metals and oxide ceramics, where the metal has a low propensity to react with O. For such weakly bonding systems, the induction of these electrostatic image charges in the metal has widely been proposed as the major bonding interaction [31–35]. Still, the possibility of a competition between the image-charge model and a moderate electron transfer and its consequences for the interface structure has raised controversial discussions [23, 24, 32, 33, 62], which were partially clarified by the inclusion of the atomistic structure of the metal with the discrete classical model (DCM) [34, 62].

Effective Many-Body Potentials

A more atomistic approach to the calculation of interface properties than the image-charge model is provided by effective pair potentials, mainly for metallic systems. Several different implementations exist and were optimised for specific applications. Generally those potentials are of the form:

$$v_{\text{eff}}(r_{ij}) = v(r_{ij}) + f(\rho). \tag{4.1}$$

The standard pair potential $v(r_{ij})$, for instance a Lennard–Jones or Morse potential, is augmented by a site-specific embedding function f, which depends on the average electron density of the crystal, $\bar{\rho}$ and can also be a function of the actual density ρ. The different approaches differ mainly in the specific form of the embedding function. To name a few popular implementations: the embedded atom model (EAM) [63, 64] employs a superposition of spherical atomic charges to build ρ, and has most successfully been applied to grain boundaries in face-centered cubic metals [64]. Finnis–Sinclair potentials are derived from TB calculations; thus they employ the square root of the density, which corresponds to the functional form of the spectral representation of the electron density [65]. Both approaches model the influence of the metallic background on the interatomic interaction. This leads to two improvements over the simple pair-potential approach: the strengthening of the remaining bonds, when the local coordination number is reduced, and the correct ordering of the elastic constants.

More complex potentials include also interactions with the third and fourth nearest neighbour atoms. These long-range interactions are, for instance, necessary to distinguish between cubic and hexagonal close-packed structures, whose first and second nearest neighbour arrangement is identical. Such an environment dependence can be introduced by a non-uniform embedding function or by the explicit addition of further interaction terms [66, 67]. In this way, also less dense-packed metals such as the body-centered cubic metals V, Nb, Mo, and W, or semiconductors such as Si can be studied. Even grain boundary structures can be analysed to some extent; yet, the delicate balance of bonding and dense-packing contributions, e. g., at the $\Sigma 5(310)[110]$ tilt grain boundary in Mo, induces a shift away from the symmetrical CSL boundary model, which could only be obtained by fully self-consistent DF calculations [68].

Ionic Models

Finally, model approaches for the calculation of the structural and elastic properties of ionic solids shall be discussed. The simplest case is the purely ionic Born model for a two-component system, which consists only of the Madelung term for the long-range Coulomb interaction between ionic point

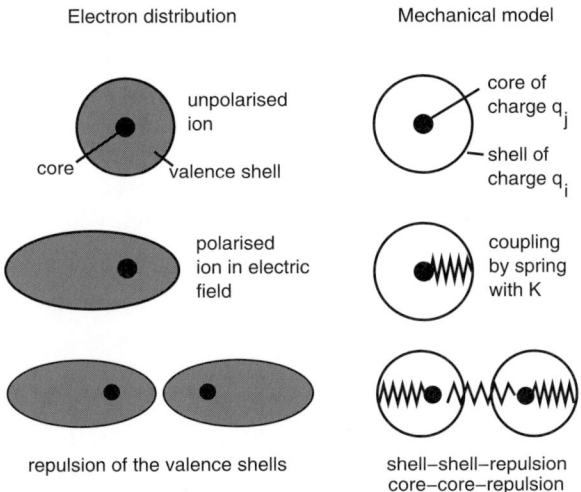

Fig. 4.7. Major interaction terms of the shell model or compressible ion model

charges and a short-range repulsive Buckingham potential to model the Pauli repulsion [69]. A modified Born model was successfully applied to the calculation of the structural data for several ionic solids [34]. In that approach the repulsive potential added to the Madelung term is free of system-specific parameters by mapping the Pauli repulsion of the ions to the isoelectronic noble gases and their van-der-Waals parameters.

The shell model is a simple mechanical analogue, which couples the ionic polarisation to the short-range pair potential [70]. Each ion is comprised of a massless shell of charge q_i, an undeformable core of charge q_j, and a harmonic spring with constant K, that couples core and shell. The sum $q_i + q_j$ gives the total charge of the ion, and the polarisability of the ion amounts to $q_i^2/2K$. Figure 4.7 sketches the major interaction types covered by the shell model.

Further extensions have been made to describe compressible ions, which account better for the deformation of the ions upon packing into the crystalline lattice. With this approach the elastic constants of several bulk oxides and halides could be correctly reproduced [34]. However, the number of adjustable parameters increases as soon as one introduces higher-order polarisation effects or as one tries to model ternary or quaternary systems. Nevertheless, the shell model was employed successfully to screen model structures of grain boundaries and surfaces of complex oxides such as α-Al$_2$O$_3$ [17] or SrTiO$_3$ [71]. In these model cases, where most structures were also accessible for a first-principles treatment, it could be verified, that the shell model yields the correct ordering with respect to the stability of the grain boundaries, and also the correct termination plane. The absolute binding energies, however, were consistently overestimated by a factor of two.

In summary, the simple classical models yield very good qualitative results for the systems they had originally been designed for. This indicates the limitations which hamper their combination to a modelling of devices composed of different material types: The development of the required coupling terms between, e.g., the shell model for ions and the embedded-atom approach for metals is not straightforward. Thus, whenever it is computationally feasible, a DF-based method will be employed to obtain quantitative results. On the other hand, an educated guess can always be derived from bulk properties on the basis of the five contributions to the bonding, which were outlined in Sect. 4.1.

4.3 Homophase Boundaries

Grain-boundary-specific structural, electronic, and elastic properties will be discussed in this chapter. The main focus is on DF investigations of oxide grain boundaries, which have been intensively studied in the recent literature [2,3,11,17,72–79]. Boundaries crucially influence the mechanical properties of *structural ceramics*, which serve e.g. as catalyst carriers, as chemically inert and wear-resistant mechanical components, or even as templates in the synthesis of photonic crystals. They also modify the electronic or optical properties of *functional ceramics*, which exhibit an additional specific functionality, such as the sensitivity to components in the chemical environment required for gas sensors, to external electric or magnetic fields provided by ferroic switches, or to temperature changes as in pyroelectric devices.

Among the main group oxides Al_2O_3 is the compound with the highest relevance for applications as a structural ceramic, thus the mechanical stability and the geometric (and also electronic) properties of its grain boundaries have been investigated in great detail, e.g. in Refs. [11,17,72–74]. Among the transition metal oxides the rutile-type TiO_2 is one of the most well-studied compounds, because it is a powerful white pigment. Thus, also grain boundaries and their structural, electronic, and optical properties have been investigated [75,76], whereas the anatase-type modification of TiO_2 became of importance only recently due to its ability to form a dilute magnetic semiconductor upon doping with Co [12]. Grain boundaries have even been studied in complex ternary oxides such as $BaTiO_3$ and $SrTiO_3$, two of the most important ferroelectric perovskite materials [2,3,77–79]. In those functional ceramics the grain-boundary-specific electronic properties are most important, because they can change the functional characteristic of the oxide and enhance or reduce its suitability for a given application. Analogous investigations have also been carried out for grain boundaries in metals such as Nb, Ta, Mo, or W [68,80,81] or in semiconductors such as GaAs [82]. The present discussion focuses on ceramic materials, therefore the reader is referred to the literature for more details on metal or semiconductor grain boundaries.

4.3.1 Pristine Boundaries

The starting point for all calculations is the setup of a supercell model, which contains the boundary plane and sufficient further lattice planes, so that the full structural relaxation of the atoms in the vicinity of the boundary can be treated. In the absence of experimental structure data one will start from the atom arrangement provided by the CSL concept and then refine the atom positions and the translation state. This concept is applicable to both *tilt* grain boundaries and *twist* grain boundaries. As depicted in Fig. 4.8 tilt grain boundaries are obtained from the CSL, if the tilt axis of the two crystals lies in the boundary plane, whereas twist grain boundaries occur, when the tilt axis is perpendicular to the interface plane.

Most commonly, three-dimensionally periodic boundary conditions are employed. They trivially extend the crystal in the two directions parallel to the grain boundary. In order to maintain translational symmetry also perpendicular to the boundary the supercell models have to contain at least two boundaries.

Figure 4.9 gives an example for the supercell construction. It shows schematically the projection of the supercell along the [1-10] tilt axis, thus columns of Ti, O, and alternating Sr and O atoms are depicted. When the

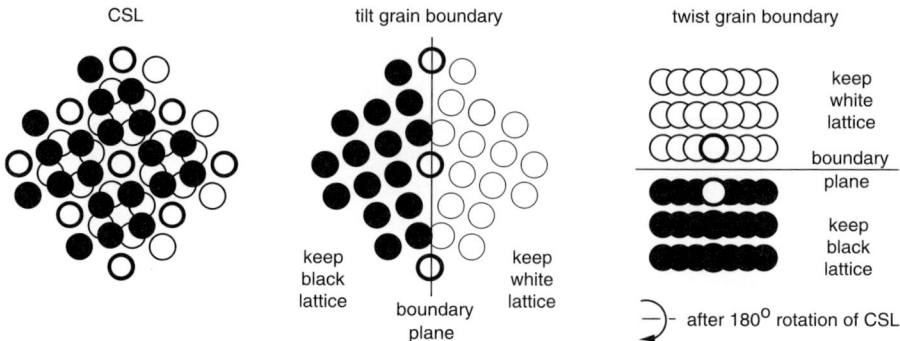

Fig. 4.8. Construction of tilt and twist grain boundaries from the CSL

atomic positions are optimised, the two very closely spaced Ti columns adjacent to the boundary plane relax away from the boundary whereas the Sr ions move in the opposite direction. Additionally, a small expansion (0.16 Å) of the supercell perpendicular to the interface was obtained, which agrees with recent experimental data within the experimental accuracy [83]. This cell is the smallest possible supercell of the $\Sigma3(111)[1\text{-}10]$SrTiO$_3$ boundary, which can be repeated periodically, and provides at least one bulk unit cell between the two boundary regions. In fact, this model is structurally identical with a less stable, hexagonal modification of SrTiO$_3$. This shows up also in the grain boundary energy E_{gb}, which is calculated as

$$E_{\mathrm{gb}} = E_{\mathrm{sc}}^{\mathrm{tot}} - n \cdot E_{\mathrm{bulk}}^{\mathrm{tot}}, \tag{4.2}$$

where $E_{\mathrm{sc}}^{\mathrm{tot}}$ is the total energy calculated for the structurally relaxed supercell, $E_{\mathrm{bulk}}^{\mathrm{tot}}$ is the total energy of one formula unit of the bulk ground state, and n is the number of formula units contained in the supercell. By this definition, E_{gb} must be a positive number, because the grain boundary is always a structural perturbation of the bulk ground state. Its upper bound is the surface energy E_{surf} (also denoted as γ in the literature), because when E_{gb} exceeds $2\,E_{\mathrm{surf}}$, it is thermodynamically more favourable for the boundary to split into its two fragments.

This occurs most likely, when the coordination environment of the ions in the boundary zone deviates strongly from the ideal bulk arrangement. This criterion, for instance, leads to the exclusion of several model structures in the screening of α-Al$_2$O$_3$ grain boundaries [17,72]. The corresponding quantity is the work of separation W_{sep}, which is obtained as

$$W_{\mathrm{sep}} = E_{\mathrm{gb}} - 2E_{\mathrm{surf}}. \tag{4.3}$$

A negative value of W_{sep} indicates bonding or adhesion, whereas for a positive value fragmentation is more favourable. A complication of this simple

argument arises, when the splitting of the grain boundary leads to two polar surfaces. Such surfaces are thermodynamically unstable and yield a high
calculated surface energy, which provides a high barrier to the fragmentation.
Experimentally, such boundaries and surfaces can release this excess energy by
facetting, reconstruction, or by the compensation of the surface polarisation
with excess and/or dopant atoms. Thus, the stability of a grain boundary may
be greatly overestimated, if the relaxation of the free surfaces is not properly
taken into account.

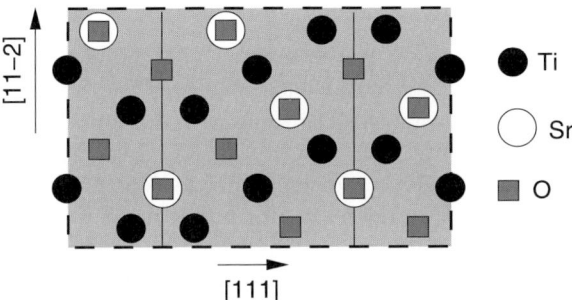

Fig. 4.9. Structure model for the two homophase boundaries $\Sigma 3(111)[1$-$10]$ of
SrTiO$_3$, derived from the CSL

Another important issue to discuss is the presence of two boundaries within
the supercell employed in the calculation, which can lead to either a stabilisation or an unfavourable interaction between the two interfaces. In the model
boundary depicted in Fig. 4.9 the spacing between adjacent grain boundaries
amounts to only 5.5 Å. Investigations with two times larger supercells have
been carried out in the meantime, e.g. in Ref. [77], which show that even
in this ionic solid the interaction decays quite rapidly and leads to an error,
which is of the order of only 10% of the grain boundary energy for the small
supercell. In agreement with experimental observations the structural relaxations are confined to a small zone close to the boundary, thus no long-range
elastic contributions arise, which would require a larger supercell model for
a proper treatment. The same is true for the modelling of space charge layers,
which have been discussed as a possible origin for deviations from the average
surface polarisability in the vicinity of grain boundaries in ferroelectric materials. Since those space charge layers extend up to several lattice constants,
they cannot be contained in the small model depicted in Fig. 4.9, neither in
the larger model of Ref. [77]. Thus, the first-principles investigations which
can be achieved with the presently available computational power, cannot significantly contribute to the issue of the extended space charge layers in the
pristine oxide boundaries.

4.3.2 Non-Stoichiometric and Doped Boundaries

The presence of interstitial atoms, vacancies, or dopant atoms at a grain boundary changes its thermodynamics. The quantity to be addressed is no longer the total energy of the boundary, but rather the grand canonical potential Ω, which contains E_{gb}, but also corrects for the non-stoichiometry via the chemical potentials μ_i of the N_C components present in the system:

$$\Omega = E_{gb} - TS - \sum_{i=1}^{N_C} \mu_i N_i . \qquad (4.4)$$

At $T = 0\,K$ the entropy term vanishes. The chemical potentials can be evaluated by calculating the total energy of the respective standard reference compound. For most crystals the reference compound is the ground-state bulk phase, but if elements such as O occur as excess atoms at the boundary, the energy of the free O_2 molecule is needed consistently with the other band-structure data. DFT, however, is not particularly well suited to compute the ground state of compounds, in which more than one Slater determinant contribute significantly to the ground-state wave function. A possible solution for this problem is the choice of a different reference compound, e.g. an oxide which contains the excess stoichiometry, e.g. SrO in the case of non-stoichiometric $SrTiO_3$ boundaries. In this case, a justification for this choice of reference comes from the experimental observation of Ruddlesden–Popper phases, which contain excess layers of SrO [84].

Grain boundary defects also modify the electric and electronic properties of a compound. Coming back to the example of the $\Sigma3(111)[1\text{-}10]SrTiO_3$ boundary the influence of the electronic structure is most obviously visualised in the plot of the spectral representation of the electron density, the density of states (DOS). The DOS of various different model supercells is compared with the bulk DOS in Fig. 4.10. Pure $SrTiO_3$ and the pristine boundaries are insulators with a band gap separating occupied and unoccupied electronic states. Regardless of the specific atom arrangement in the vicinity of the grain boundary, the Fermi level is located at the band gap onset in all stoichiometric models, thus they are electronically inactive. If O vacancies are present either in the boundary plane ($V(O_0)$) or close to the boundary ($V(O_2)$), the Fermi level lies within the manifold of metal-derived states at the conduction band edge, whereas doping of the grain boundary with trivalent metal atoms formally oxidises the O atoms in that region, thus a more complex electron rearrangement takes place. In both cases, the presence of the defects at the grain boundary provides electronic states at the Fermi level, which indicate an electronic contribution to the conductivity of $SrTiO_3$. From the energy differences between the vacancy-containing models $V(O_0)$ and $V(O_2)$ one can even conclude that the grain boundary acts as a sink for O vacancies, thus it also influences the vacancy mobility, which makes up the major part of the conductivity of $SrTiO_3$ [3].

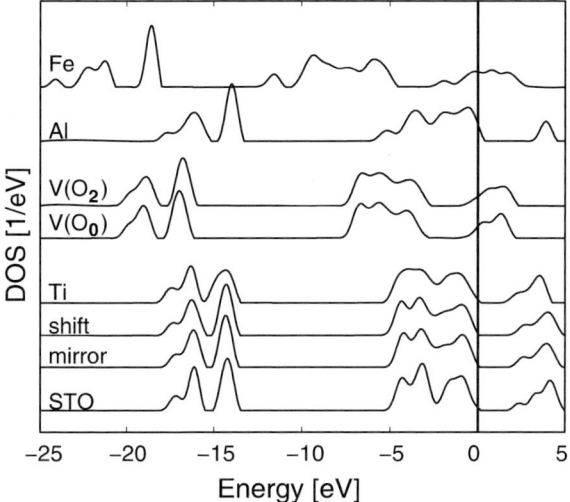

Fig. 4.10. Total DOS of bulk crystalline $SrTiO_3$, the mirror-symmetric and the shifted SrO_3-terminated $\Sigma 3(111)[1\text{-}10]SrTiO_3$ boundary, a Ti-terminated model, two vacancy-containing supercells, and the Al-doped and Fe-doped boundaries from *bottom* to *top*. For a better comparison the DOS curves are shifted along the DOS axis, and the Fermi level is set to zero

Finally, impurities also influence the macroscopic elastic behaviour of a compound by grain boundary embrittlement. Famous examples are the grain boundary embrittlement of Cu by Bi or of Al alloys by Ga. Recent first-principles modelling has shown that for Bi in Cu the large size of the Bi atom leads to undercoordinations at the grain boundary once a complete Bi bilayer is formed, and to subsequent mechanical failure of the material [85]. For Ga in Al, however, also electronic factors come into play [86].

4.4 Heterophase Boundaries

At heterophase boundaries, where two different materials come into contact with each other an even richer variety of boundary-related phenomena is observed than at homophase boundaries. This chapter focuses on metal–ceramic and metal-semiconductor interfaces, where the electronic structures of the components differ most strongly, and a first-principles electronic structure theory can best demonstrate its modelling power. Nevertheless, it has proven its applicability also in numerous investigations of semiconductor-insulator boundaries such as C on Si [87], Si on SiO_2 [88], TiN/MgO(001) [89], or ZrO_2/Si and $ZrSiO_4$/Si [90], to name some examples from the literature.

Apart from boundaries between two different material types, also the properties of junctions between two different metals, semiconductors, or insulators have been studied. The investigations of strain measurements at the Ag/Ni(001) interface [91], the comparative adhesion of Pd or Ag on the Nb(001) and Nb(110) faces [92], and further theoretical studies of thin film and multilayer systems [93], such as Ni/Ru(0001) [94], Pt/Ta [95], Pd/Nb(110) [96], Cu/Ta [97], or Ag/Pt [98] are examples, partly motivated by the observations of giant magnetoresistance in metallic multilayers. Also the III–V and II–VI semiconductors provide ample possibilities to tune specific optical properties by producing multilayers and adjusting the electronic and elastic properties by compositional variation. Further examples include the III–V combinations GaAs/AlAs [99, 100], or GaSb/InAs(100) [100], InAs/GaSb [101], the II–VI combination CdTe/HgTe [102], and mixed multilayers of III–V and II–VI semiconductors like GaAs/ZnSe(001) [103], or of III–V on IV materials, such as SiC/AlN and SiC/BP [104], AlN/SiC(10-10) [105], or Ge/GaAs(100) [106]. In these and other more covalently bonded systems, like SiC/Si(001) [107], the main focus of the investigations is on the interplay between lattice strain and electronic properties. Other insulator–insulator boundaries come from the heteroepitaxy of ferroic oxides on an oxide template, such as $SrTiO_3/MgO$ [108], or multilayers such as $(Ba_{1/3}Sr_{1/3}Ca_{1/3})TiO_3$ [109] or $BaTiO_3/SrTiO_3$ [110].

4.4.1 Wetting and Growth of Metal Layers

A vital interest in a microscopic understanding of the interactions at metal–ceramic interfaces has developed over the last decade, motivated by the increasing use of ceramic materials in various industrial applications, for instance as sensor materials [111], as thermal barrier coatings [112], or even as implants for medical purposes [113]. This interest has challenged various attempts for theoretical modelling of the relevant contributions which influence the bonding behaviour at the interface. Recent studies span the whole range from finite-element modelling to understand elastic properties [114], over atomistic simulations of dislocation networks, plasticity, and fracture [115], to the investigation of the electronic structure with ab-initio band-structure techniques.

From the theoretical modelling and the experimental observations on metal–ceramic bonding, three scenarios can be distinguished:

(1) Strong adhesion is found for main-group or early transition metals on oxides with a propensity of the metal to bind on top of the O atom. This behaviour is, for instance, observed in Ti on MgO(100) [23], for Al on $Al_2O_3(0001)$ [26], for V on MgO(001) [116], or for Al and Ti on $MgAl_2O_4$ (001) [4–9].

(2) For late transition metals only weak adhesion was found at the non-polar oxide surfaces for instance for Ag/MgO(001) [18,23,117], for Cu/MgO(001) [115], for the $VO_x/Pd(111)$ interface [118], or for Ag on $MgAl_2O_4(001)$ [4, 6, 9]. Due to the also experimentally apparent lack of strong bond-

ing [19], the underlying adhesion mechanism was ascribed to image charge interactions [31–35].

(3) A moderately strong bonding is obtained, when additional elastic interactions interfere with the metal-to-oxygen bonding. For the adhesion of Cu on the polar (111) surface of MgO, however, evidence for a direct metal-oxygen interaction was given and the occurrence of metal-induced gap states was postulated [119]. Presently, transition-metal oxides such as $BaTiO_3$ or $SrTiO_3$ are studied as substrates for the adhesion of transition metals like Pd, Pt, or Mo [120] or $ZnO(0001)/Pd(111)$ [121]. Similar adhesive interactions are also monitored for metal-insulator contacts, where the insulator is not an oxide, but a carbide [122, 123] or nitride [124]. In addition to band structure calculations, cluster methods have been employed frequently to study the adhesion properties of small clusters or the first nucleation steps of film growth [125].

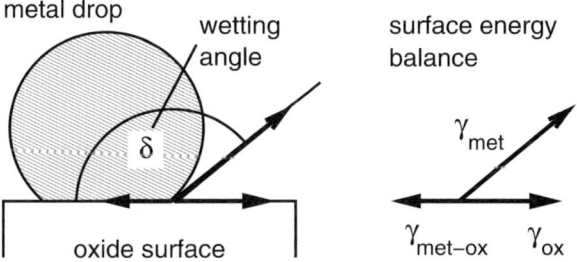

Fig. 4.11. Definition of the wetting angle and the balance of the surface tensions

From a phenomenological point of view, the three regimes can be characterised by the growth mode during metal deposition and by the so-called wetting angle. Figure 4.11 depicts the definition of the wetting angle θ and its correlation to the surface tensions γ_{ox} and γ_{met} of oxide and metal, and to the interface energy γ_{met-ox}. Assuming a balance of the surface tensions at the triple point, where the two surfaces and the interface meet, the connection between those quantities is given by Young's equation [126]:

$$\gamma_{ox} - \gamma_{met-ox} = \gamma_{met} \cdot \cos(\theta). \tag{4.5}$$

If the interface is more stable than the sum of the two free, relaxed surfaces complete wetting of the ceramic by a (thin) metal film is achieved, and a low wetting angle $\theta \approx 0°$ obtained. In this case, the metal can be deposited in a layer-by-layer growth mode of Franck-van-der-Merwe type (see Fig. 4.12).

On the contrary, if the free surfaces are more stable than the interface, the metal will form droplets in order to minimise the size of the unfavourable interface with the ceramic surface. The wetting angle will approach 180°, and a Volmer–Weber island growth corresponds to this situation. As discussed for homophase boundaries, it has to be considered, that the calculated energy of an unrelaxed polar ceramic surface may be considerably higher than the experimentally observable one of a structurally relaxed surface. For intermediate angles around the wetting–dewetting transition point with $\theta \approx 90°$, the growth mode depends sensitively on the deposition conditions. This phenomenon should, however, not be confused with the mixed-mode growth of islands on a thin wetting layer (Stranski–Krastanov growth), which is a completely stress-driven relaxation of an otherwise well wetting system. For an exhaustive description of the nucleation and growth processes in thin-film formation the reader is referred to the review article by Venables et al. [127]. Of the three adhesion regimes outlined above, case (1) may be most closely correlated to complete wetting of the oxide surface by the metal, whereas case (2) resembles more the dewetted situation, and case (3) may be in the region, where the dewetting transition occurs.

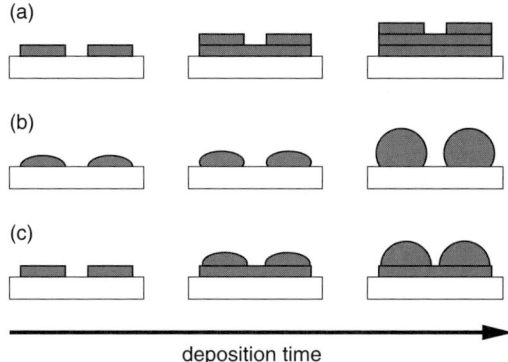

Fig. 4.12. Growth modes for metals on ceramic surfaces: (**a**) Franck-van-der-Merwe-type layer-by-layer growth, (**b**) Volmer–Weber island growth, and (**c**) misfit-induced Stranski–Krastanov growth of islands on a thin wetting layer

4.4.2 Metal–Ceramic Boundaries

As mentioned above for the intermediate wetting case (3) it is often difficult to compare theoretical predictions with experimental results, since the mismatch in lattice constants between the metal and the ceramic gives rise to periodically repeated misfit dislocations in the metal [18, 19]. These dislocations influence other observed properties, such as the wetting angle, but were not accounted for in the idealized theoretical models. In order to avoid these complications two model systems were studied by one of the authors, which have almost no lattice mismatch (Al/MgAl$_2$O$_4$ with 0.27% mismatch and Ag/MgAl$_2$O$_4$ with 1.25% mismatch). They hence meet the

basic requirement for the formation of a coherent interface with a small
unit cell, which is crucial for the theoretical description of a realistic system. Furthermore, the choice of a main-group metal and a noble metal provides the possibility to compare the mechanisms of both strong and weak
bonding to the same substrate. From a theoretical point of view this is
especially interesting since the valence electronic structures of Al atoms
(configuration $s^2 p^1$) and Ag atoms $(d^{10} s^1)$ exhibit a formal similarity with
one unpaired electron per atom. The publications [7–9] deal with this adhesive material combination, which was the first system that has been studied systematically and in great detail both by theoretical and experimental
investigations.

The supercells for the study of the Al/spinel and Ag/spinel boundaries are
depicted schematically in Fig. 4.13. The relative translation state was calculated by optimisation both of the atom positions and the supercell dimensions
for the two different terminations of the spinel slab by either a dense-packed
layer of AlO_2 stoichiometry or by a sparse Mg layer.

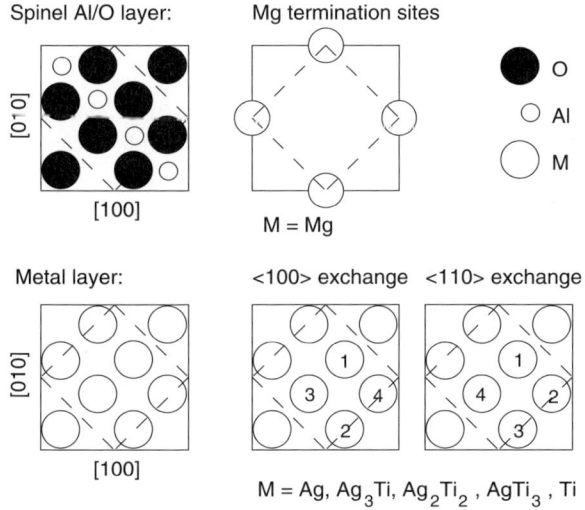

Fig. 4.13. Supercell model for the adhesion of Al(001) and Ag(001) on the (001)
surface of spinel. The numbers denote the order of the stepwise exchange of Ag
atoms by Ti atoms in a detailed analysis of the Ti-induced adhesion enhancement

The DF band-structure calculations indicate that the termination of the
spinel by a layer composed of Al and O ions is favored over a termination
by a layer sparsely occupied by Mg ions for the following reasons: Very good
agreement with results of high-resolution transmission electron microscopy experiments [128] is obtained for the $Al/MgAl_2O_4$ system, with a well-defined
Al/O adhesion site on top of the O ions of the spinel at an interface spacing

of $1.90\,\text{Å}$ close to the experimental one of $1.94 \pm 0.06\,\text{Å}$. For Mg termination, however, a stretched interface distance of $2.45\,\text{Å}$ was calculated for the Al/O adhesion site, in disagreement with experiment [128]. With a work of separation of $2.25\,\text{J/m}^2$ the position on top of the O ions is considerably more stable than the positions above the three-fold and four-fold coordinated hollow sites $\left(W_{\text{sep}} = 1.1\,\text{J/m}^2\right)$ or the position on top of Mg $\left(W_{\text{sep}} = 0.6\,\text{J/m}^2\right)$.

For the $Ag/MgAl_2O_4$ system any translation state on top of the O ions or on the hollow sites of the spinel can be occupied by the Ag atoms with similar energies. This result may explain the experimental observation that the cohesion between Ag and spinel is very weak and Ag prefers the formation of the thermodynamically more stable (111) surface. The M/Al site can be ruled out both for Al and Ag adhesion because of the much higher interface energies compared to the M/O sites.

An analysis of the electronic structure yields more insight into the different interaction mechanisms. For this purpose the electron density redistribution is analysed, which occurs when the interface is formed from the two free surfaces, because this measure can be related to the interaction to be overcome upon fracture. For this purpose the valence electron densities in a plane perpendicular to the surface or interface were calculated for each of the free surfaces and the interface, separately, and then the densities for the surfaces subtracted from the density for the interface. The corresponding density difference plots on a (010) plane of the cubic lattice are depicted in Fig. 4.14, where dark areas indicate charge depletion and light ones charge accumulation.

The density difference shown in the left panel is calculated for $Al/MgAl_2O_4$ in the arrangement Al on top of O, which is the low-energy adhesion geometry. Accordingly for $Ag/MgAl_2O_4$ the adhesion at the hollow-site position is analysed as depicted in the right panel of Fig. 4.14. In both density difference plots the metals are located in the center part of the boundary model and spinel in the upper and lower parts as indicated in the plots. The two interfaces in the repeat unit cross the graphs horizontally in the upper and lower parts. In the Al/spinel system the main contribution to bonding is the charge redistribution between the metal atoms and the oxygen ions. Also some minor lateral charge rearrangements inside the spinel and the metal are induced. In the Ag/spinel system a structureless charge flow from the spinel, some rehybridization of the Ag atoms, and a subsequent strong polarization of the metal sheet are obtained from calculation. Even at the more electron-rich boundary, where Ag is placed at the three-fold coordinated hollow sites, no significant bonding is observed theoretically. These findings are indicative of a strong Pauli repulsion between the localised Ag-d orbitals and the closed shells of the oxide ions. Also the higher lattice misfit of Ag/spinel compared with Al/spinel adds unfavourably to the metal–ceramic interaction.

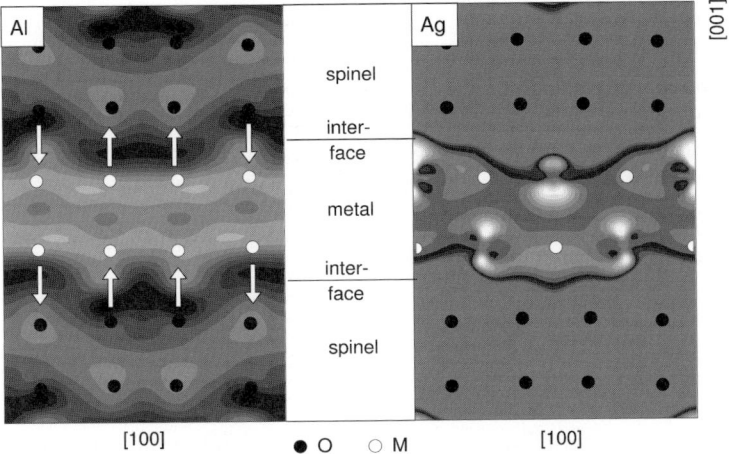

Fig. 4.14. Valence charge density differences for Al/spinel (*left*) and Ag/spinel (*right*). The positions of atoms neighboring the cut plane are indicated schematically by white circles for the metals and black circles for the O atoms of spinel; the Mg and Al ion positions in spinel are omitted for clarity. The plots depict more than one repeat unit along the [001] direction, such that the full spinel layer is visible both above and below the metal layer and that the next periodic replicas of the two interfaces are visible at the top and the bottom of the plot. Al bonds to the M/O adhesion site, Ag is coordinated to hollow site positions formed by two O ions and one Al ion. The arrows in the left panel indicate the complex electron rearrangement pattern over the boundary, which leads to the bonding

In weakly bonding systems, however, the adhesion can be enhanced by the insertion of a reactive interlayer. In the present case Ti was chosen, because it allows for a better discrimination of the three interaction-determining factors: charge transfer, Pauli repulsion, and elastic contributions. The relative ordering of the three metals Ti, Al, and Ag is as follows:

- according to the electronegativity: Ti < Al < Ag
- according to the electron gas parameter: Al $\approx$ Ti > Ag
- according to the lattice mismatch: Al < Ag < Ti.

With calculated interface distances of $d(\text{Al-O}) = 1.9\,\text{Å} < d(\text{Ti-O}) = 1.99\,\text{Å} < d(\text{Ag-O}) = 2.3\,\text{Å}$ and values for the work of separation of $W_{\text{sep}}(\text{Al/Sp}) = 2.25\,\text{J/m}^2 > W_{\text{sep}}(\text{Ti/Sp}) = 1.81\,\text{J/m}^2 > W_{\text{sep}}(\text{Ag/Sp}) = 1.10\,\text{J/m}^2$, the Ti-containing adhesion system exhibits clearly an average position between the two pristine boundaries. It is thus concluded, that the reduction of the Pauli repulsion by the low valence electron density of the Ti film is the main driving force for the adhesion enhancement (*titanium effect*), whereas the other two components balance. From the stepwise exchange of Ag atoms by Ti atoms within the interlayer, indicated in Fig. 4.13, it was shown, that the major stabilisation is already reached, when every second Ag atom is replaced by Ti.

One major drawback of reactive interlayers is described in Refs. [8, 9]: Early transition metals such as Ti have a high oxygen affinity. This, of course, assists the bonding with the oxide substrate, it however degrades the long-term stability of the interface in an oxidative environment.

4.4.3 Reactive Metal–Semiconductor Interfaces

Many examples for reactive interfaces are also found in the material combination occuring at metal–semiconductor contacts. Almost all early transition and rare earth metals bind to the Si(111) or Si(110) surfaces under the formation of binary and ternary silicides. The best studied example is the interface Co/Si, where DF modelling has shown the thermodynamic driving force for the formation of silicides like $CoSi_2$ [27–29, 129, 130]. The first steps of the silicide formation are the migration of Co atoms to interstitial sites in the Si lattice, the diffusion of those atoms along Ti grain boundaries, and the growth of larger $CoSi_2$ crystals from $CoSi_2$ seeds [27]. Also the reconstruction at the interface could be explained within a DF framework [28]. Slab calculations of $CoSi_2(100)/Si(100)$ [130], $CoSi_2(111)/Si(111)$[129], and of up to two monolayers of Co on Si(001) [29] confirm that the formation of the reactive oxide is not dependent on a particular orientation of the substrate. Analogous electronic structure calculations have also been carried out for Y and Gd [131], Sr [132], Ni [133] and Fe [134].

Titanium also forms several stable silicides at the boundary, which lie within a composition range of Ti_3Si to $TiSi_2$. Yet, the corresponding Ti/Si interface had so far defied a detailed treatment at the quantum mechanical level, because the lattice mismatch between the atoms in the Ti(0001) and the Si(111) planes amounts to 25%. Thus, an interface model with low remanent elastic stress can be constructed, but at the cost of a high number of atoms in the supercell. Due to the lattice mismatch between Ti ($a_0 = 2.95\,\text{Å}$) and Si ($a_0 = 3.85\,\text{Å}$) equal areas of the Ti(0001) and Si(111) surfaces contain different numbers of atoms with a Ti:Si ratio of approximately 16:9. Thus the supercells employed in the calculation span $4a_0(\text{Ti}) \approx 3a_0(\text{Si})$ parallel to the interface, and contain five layers of Ti and eight layers of Si along the direction perpendicular to the interface. The interface energy W_{sep} with respect to fracture into the two unrelaxed surfaces amounts to only $0.28\,\text{J/m}^2$, which indicates a weak adhesion. In a smaller supercell model with only $3a_0(\text{Ti}) \approx 2a_0(\text{Si})$ parallel to the interface the remaining tensile and compressive stresses make the system even non-bonding.

The geometry optimisation of Ti(0001)/Si(111) yields a rumpling of the Ti plane adjacent to the interface by 0.09 Å. By this distortion the arrangement of Ti on top of a Si atom and the position of Ti above a three-fold coordinated hole of the Si(111) surface lead to comparable Ti–Si distances in the range of

2.6 to 2.7 Å. The Si layer adjacent to the interface does not exhibit any significant rumpling, but the distance to the next Si layer is diminished by 0.21 Å, indicative of the Ti–Si attraction. The electron density difference between the interface and a superposition of the Ti and Si slabs is depicted in Fig. 4.15. It visualises that the Ti layer is little affected by the interface, and complete screening occurs in the center of the metal slab. In the semiconducting Si the screening is less efficient, and the electron accumulation in the plane adjacent to the interface illustrates the tendency for silicide formation. The asymmetry of the electron accumulation patterns in Fig. 4.15 is a consequence of the complex structural relaxations at the interface, which make the two interfaces asymmetric.

Finally, molecular-dynamics calculations show that an amorphous silicide film is accessible at temperatures, for which the crystalline binary compounds do not yet melt. The interface reaction is accompanied by an equilibration and reduction of misfit stresses and an increase of the interface energy of up to 0.52 J/m². Thus, the silicide formation is not only driven by the electronegativity difference of Ti and Si, but enhanced by an elastic contribution.

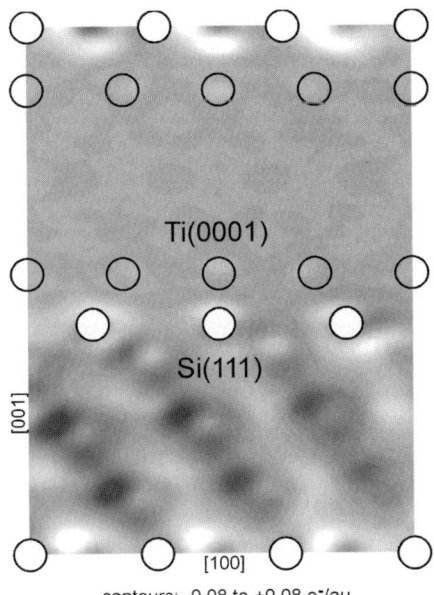

contours: -0.08 to +0.08 e⁻/au

Fig. 4.15. Binding electron density plot of the Ti/Si system in a plane perpendicular to the interface, where the on-top arrangement of Ti and Si atoms leads to Ti–Si bonds. Depicted is the difference between the electron density of the interface and the superposition of the electron densities of the two separate Ti(0001) and Si(111) slabs calculated in the same supercell. 17 grey levels span from -0.08 to $+0.08$ electrons/bohr². *Dark areas* correspond to electron depletion, *bright ones* depict electron accumulation

4.5 Summary and Outlook

The atomistic and electronic structure at planar grain boundaries and heterophase interfaces between different materials were studied in order to develop simple criteria, which allow us to predict the interface properties on the basis of known bulk properties. Five dominant interactions were found to govern the structural and electronic properties of the boundaries, and mapped to properties of the single atom or the bulk:

(1) **Coulomb** interaction between **charges** in ionic crystals,
(2) **elastic** interactions due to **misfit** strains at the interface,
(3) **electron transfer** processes across the interface, when the atoms at the interface exhibit different **electronegativity**,
(4) **Pauli repulsion** between filled valence shells with a low **electron gas parameter**,
(5) **image charge** interactions between fixed (ionic) charges and **induced dipoles** at the other side of the boundary.

Not all interactions are of equal importance: Depending on the two materials that come into contact, one or more interactions may even be absent. At grain boundaries, electron transfer or image charge interactions are negligible compared with the dominant Coulomb term. Heterophase boundaries exhibit a much richer variety of interactions, which have been analysed in detail, also by comparison between theoretical and experimental data. Table 4.1 provides a qualitative overview of the correlation between the materials present at the interface and the relevant interactions.

The first two columns denote the type of material combination and the other five columns contain the interaction contributions labelled as described in the caption. The first three rows denote the homophase boundaries, the next three rows contain heterophase systems, where the two bonding partners come from the same material class, and the last three rows contain boundaries between different materials.

Some trends can be seen at first glance: as described above, the number of relevant interactions increases with increasing complexity of the boundary from the top to the bottom of the table. Second, some parameters may become dominant only in very specific cases, e.g. the image charge interaction specific for metal-insulator boundaries, or the Coulomb interaction at the boundary between two ionic solids. The probably most important insight conveyed by the table is that elastic interactions are nearly ubiquitous, and one can avoid them only by the proper choice of systems with matching lattice constants, or fix them to some extent by doping. This comparison confirms that the fine and detailed interaction balance of the criteria (3) to (5) at the metal-spinel boundaries discussed in Sect. 4.4 is only observable due to the absence of strain.

Table 4.1. Correlation between the different contributions to interfacial bonding and the different material combinations of crystals 1 and 2. The first two columns denote the type of material: M = metal, S = semiconductor (and, for larger gaps, covalent insulator), and I = ionic insulator. The entry "++" indicates, that the interaction can become very important, "+" means that the interaction can be relevant, and "o" is chosen for cases, in which the interaction typically is not relevant

crystal 1	crystal 2	Coulomb	elastic	electron transfer	Pauli repulsion	image charges
M	M	o	++	+	o	o
S	S	o	++	o	o	o
I	I	++	+	o	o	o
M	M'	o	+	++	+	o
S	S'	+	++	+	+	o
I	I'	++	++	++	o	o
S	I	+	++	+	o	o
M	S	o	++	+	o	+
M	I	o	++	+	++	++

References

1. A.P. Sutton, R.W. Balluffi: *Interfaces in Crystalline Materials* (Clarendon Press, Oxford 1995)
2. S. Hutt, S. Köstlmeier, C. Elsässer: J. Phys.: Condens. Matter **13**, 3949 (2001)
3. S. Gemming, M. Schreiber: Chem. Phys. **309**, 3 (2005)
4. S. Köstlmeier, C. Elsässer: Interface Science **8**, 41 (2000)
5. S. Köstlmeier, C. Elsässer, B. Meyer, M.W. Finnis: Mat. Res. Soc. Symp. Proc. **492**, 97 (1998)
6. S. Köstlmeier, C. Elsässer, B. Meyer, M.W. Finnis: phys. stat. sol. (a) **166**, 417 (1998)
7. S. Köstlmeier, C. Elsässer: J. Phys.: Condens. Matter **12**, 1209 (2000)
8. C. Elsässer, S. Köstlmeier-Gemming: Phys. Chem. Chem. Phys. **3**, 5140 (2001)
9. S. Köstlmeier, C. Elsässer: Mat. Res. Soc. Symp. Proc. **586**, M3.1 (1999)
10. I. Levin, D. Brandon: J. Am. Ceram. Soc. **81**, 1995 (1998)
11. S. Nufer, A.G. Marinopoulos, T. Gemming, W. Kurtz, C. Elsässer, S. Köstlmeier, M. Rühle: Phys. Rev. Lett. **86**, 5066 (2001)
12. P.A. Stampe, R.J. Kennedy, Y. Xin, J.S. Parker, J. Appl. Phys. **93**, 7864 (2003)
13. R. Schweinfest, S. Köstlmeier, F. Ernst, C. Elsässer, T. Wagner, M.W. Finnis: Phil. Mag. A **81**, 927 (2000)
14. S. Köstlmeier, C. Elsässer, B. Meyer: Ultramicrosc. **80**, 145 (1999)
15. R.F.W. Bader: *Intern. Series of Monographs on Chemistry 22* (Clarendon Press, Oxford 1990)
16. X. Gonze: Phys. Rev. B **55**, 10355 (1997)
17. A.G. Marinopoulos, S. Nufer, C. Elsässer: Phys. Rev. B **63**, 165112 (2001)

18. V. Vitek, G. Gutekunst, J. Mayer, M. Rühle: Phil. Mag. A **71**, 1219 (1996)
19. A. Trampert, F. Ernst, C.P. Flynn, H.F. Fischmeister and M. Rühle: Acta Metall. Mater. **40**, S227 (1992)
20. S.J. Bennison, M.P. Harmer: A History of the Role of MgO in the Sintering of α-Al$_2$O$_3$, in: *Sintering of Advanced Ceramics*, ed. by C.A. Handwerker, J.E. Blendell (The American Ceramic Society, Westerville OH 1989)
21. P.O. Hanson, F. Ernst: J. Appl. Phys. **72**, 2083 (1992)
22. J. Bardeen: Phys. Rev. **71**, 717 (1947)
23. U. Schoenberger, O.K. Andersen, M. Methfessel: Acta Metall. Mater. **40**, S1 (1992)
24. J.R. Smith, T. Hong, D.J. Srolovitz: Phys. Rev. Lett. **72**, 4021 (1994)
25. C. Li, R. Wu, A.J. Freeman, C.L. Fu: Phys. Rev. B **48**, 8317 (1993)
26. K. Kruse, M.W. Finnis, J.S. Lin, M.C. Payne, V.Y. Milman, A. DeVita, M.J. Gillan: Phil. Mag. Lett. **73**, 377 (1996)
27. A. Horsfield, H. Fujitani: Phys. Rev. B **63**, 235303 (2001)
28. B.D. Yu, Y. Miyamoto, O. Sugino, A. Sakai, T. Sasaki, T. Ohno: J. Vac. Sci. Technol. B **19**, 1180 (2001)
29. B.S. Kang, S.K. Oh, H.J. Kang, K.S. Sohn: J. Phys.: Condens. Matter **15**, 67 (2003)
30. A. Szabo, N.S. Ostlund: *Modern Quantum Chemistry, Introduction to Advanced Electronic Structure Theory* (McGraw – Hill, New York 1989)
31. A.M. Stoneham, P.W. Tasker: J. Phys. C: Solid State Phys. **18**, L543 (1985)
32. D.M. Duffy, J.H. Harding, A.M. Stoneham: Acta Metall. Mater. **40**, S11 (1992)
33. D.M. Duffy, J.H. Harding, A.M. Stoneham: Phil. Mag. A **67**, 865 (1993)
34. M.W. Finnis: Acta Metall. Mater. **40**, S25 (1992)
35. A.M. Stoneham, P.W. Tasker: Phil. Mag. B **55**, 237 (1987)
36. R.O. Jones, O. Gunnarsson: Rev. Mod. Phys. **61**, 689 (1989)
37. H. Eschrig: *The Fundamentals of Density Functional Theory* (Edition am Gutenbergplatz, Leipzig 2003)
38. G. Onida, L. Reining, A. Rubio: Rev. Mod. Phys. **74**, 601 (2002)
39. R.M. Dreizsler, E.K.U. Gross: *Density Functional Theory* (Springer, Berlin 1990)
40. R.G. Parr, W. Yang: *Density-Functional Theory of Atoms and Molecules* (Oxford University Press, New York 1989)
41. A.M. Stoneham, J.H. Harding: Nature Materials **2**, 65 (2003)
42. M.W. Finnis: *Interatomic Forces in Condensed Matter* (Oxford University Press, Oxford 2003)
43. P. Hohenberg, W. Kohn: Phys. Rev. **136**, B864 (1964)
44. M. Levy: Phys. Rev. A **26**, 1200 (1982)
45. U. von Barth, L. Hedin: J. Phys. C **5**, 1629 (1972)
46. N.D. Mermin: Phys. Rev. **137**, A1441 (1965)
47. S.H. Vosko, L. Wilk, M. Nusair: Can. J. Phys. **58**, 1200 (1980)
48. O. Gunnarson, M. Jonson, B.I. Lundqvist: Phys. Rev. B **20**, 3136 (1979)
49. R. Car, M. Parrinello: Phys. Rev. Lett. **55**, 2471 (1985)
50. N.W. Ashcroft, N.D. Mermin: *Solid State Physics* (Saunders College, Philadelphia PA 1976)
51. J.-M. Jancu, R. Scholz, F. Beltram, F. Bassani: Phys. Rev. B **57**, 6493 (1998)
52. R. Scholz, J.-M. Jancu, F. Bassani: Mat. Res. Soc. Symp. Proc. **491**, 383 (1998)

53. A. Di Carlo: Mat. Res. Soc. Symp. Proc. **491**, 391 (1998)
54. C.Z. Wang, K.M. Ho, C.T. Chan: Phys. Rev. Lett. **70**, 611 (1993)
55. P. Ordejón, D. Lebedenko, M. Menon: Phys. Rev. B **50**, 5645 (1994)
56. M. Menon, K.R. Subbaswamy: Phys. Rev. B **55**, 9231 (1997)
57. C.M. Goringe, D.R. Bowler, E. Hernandez: Rep. Prog. Phys. **60**, 1447 (1997)
58. T.A. Niehaus, S. Suhai, F. Della Sala, P. Lugli, M. Elstner, G. Seifert, T. Frauenheim: Phys. Rev. B **63**, 085108 (2001)
59. A. Di Carlo, M. Gheorghe, P. Lugli, M. Sternberg, G. Seifert, T. Frauenheim: Physica B **314**, 86 (2002)
60. N.D. Lang, W. Kohn: Phys. Rev. B **7**, 3541 (1973)
61. N.V. Smith, C.T. Chen, M. Weinert: Phys. Rev. B **40**, 7565 (1989)
62. J. Purton, S.C. Parker, D.W. Bullett: J. Phys.: Condens. Matter **9**, 5709 (1997)
63. M.S. Daw, M.I. Baskes: Phys. Rev. Lett. **50**, 1285 (1983)
64. M.S. Daw, M.I. Baskes: Phys. Rev. B **29**, 6443 (1984)
65. M.W. Finnis, J.E. Sinclair: Phil. Mag. A **50**, 45 (1984)
66. J.A. Moriarty, R. Phillips: Phys. Rev. Lett. **66**, 3036 (1990)
67. J. Tersoff: Phys. Rev. B **38**, 9902 (1988)
68. T. Ochs, O. Beck, C. Elsässer, B. Meyer: Phil. Mag. A **80**, 351 (2000)
69. J.H. Harding: Rep. Prog. Phys. **53**, 1403 (1990)
70. B.G. Dick, A.W. Overhauser: Phys. Rev. **112**, 90 (1958)
71. E. Heifets, E.A. Kotomin, R. Orlando: J. Phys.: Condens. Matter **8**, 6577 (1996)
72. C. Elsässer, A.G. Marinopoulos: Acta Mater. **49**, 2951 (2001)
73. A.G. Marinopoulos, C. Elsässer: Acta Mater. **48**, 4375 (2000)
74. S. Fabris, A. Nufer, C. Elsässer, T. Gemming: Phys. Rev. B **66**, 155415 (2002)
75. I. Dawson, P.D. Bristowe, M.-H. Lee, M.C. Payne, M.D. Segall, J.A. White: Phys. Rev. B **54**, 13727 (1996)
76. S.B. Sinnott, R.F. Wood, S.J. Pennycook: Phys. Rev. B **61**, 15645 (2000)
77. R. Astala, P.D. Bristowe: J. Phys.: Condens. Matter **14**, L149 (2002)
78. R. Astala, P.D. Bristowe: J. Phys.: Condens. Matter **14**, 13635 (2002)
79. R. Astala, P.D. Bristowe: J. Phys.: Condens. Matter **14**, 6455 (2002)
80. R. Janisch, C. Elsässer: Phys. Rev. B **67**, 224101 (2003)
81. G.-H. Lu, S. Deng, T. Wang: Phys. Rev. B **69**, 134106 (2004)
82. W.R.L. Lambrecht, C. Amador, B. Segall: Phys. Rev. Lett. **68**, 1363 (1992)
83. Z. Zhang, W. Sigle, F. Phillipp, M. Rühle: Science **302**, 846 (2003)
84. S. Sturm, A. Recnik, C. Scheu, M. Ceh: J. Mater. Res. **15**, 2131 (2000)
85. R. Siegl, M. Yan, V. Vitek: Mod. Simul. Mat. Sci. Eng. **5**, 105 (1997)
86. D.I. Thomson, V. Heine, M.C. Payne, N. Marzari, M.W. Finnis: Acta Mater. **48**, 3623 (2000)
87. M. Sternberg, W.R.L. Lambrecht, T. Frauenheim: Phys. Rev. B **56**, 1568 (1997)
88. T. Sakurai, T. Sugano: J. Appl. Phys. **52**, 2889 (1981)
89. D. Chen, X.L. Ma, Y.M. Wang, L. Chen: Phys. Rev. B **69**, 155401 (2004)
90. R. Puthenkovilakam, E.A. Carter, J.P. Chang: Phys. Rev. B **69**, 155329 (2004)
91. J.A. Floro, C.V. Thompson, R. Carel, P.D. Bristowe: J. Mater. Res. **9**, 2411 (1994)
92. M. Weinert, R.E. Watson, J.W. Davenport, G.W. Fernando: Phys. Rev. B **39**, 12585 (1989)
93. R.C. Longo, V.S. Stepanyuk, W. Hegert, A. Vega, L.J. Gallego, J. Kirschner: Phys. Rev. B **69**, 073406 (2004)

94. J.E. Houston, J.M. White, P.J. Feibelman, D.R. Hamann: Phys. Rev. B **38**, 12164 (1988)
95. R.E. Watson, M. Weinert, J.W. Davenport: Phys. Rev. B **35**, 9284 (1987)
96. V. Kumar, K.H. Bennemann: Phys. Rev. B **28**, 3138 (1983)
97. H.R. Gong, B.X. Liu: Appl. Phys. Lett. **83**, 4515 (2003)
98. S. Narasimham: Phys. Rev. B **69**, 045425 (2004)
99. P.J. Lin-Chung, T.L. Reinecke: J. Vac. Sci. Technol. **19**, 443 (1981)
100. S. Das Sarma, A. Madhukar: J. Vac. Sci. Technol. **19**, 447 (1981)
101. Y. Wei, M. Razeghi: Phys. Rev. B **69**, 085316 (2004)
102. J.N. Schulman, T.C. McGill: J. Vac. Sci. Technol. **16**, 1513 (1979)
103. A. Kley, J. Neugebauer: Phys. Rev. B **50**, 8616 (1994)
104. W.R.L. Lambrecht, B. Segall: Phys. Rev. B **43**, 7070 (1991)
105. R. Di Felice, J.E. Northrup: Phys. Rev. B **56**, 9213 (1997)
106. K. Kunc, R.M. Martin: Phys. Rev. B **24**, 3445 (1981)
107. L. Pizzagalli, G. Cicero, A. Catellani: Phys. Rev. B **68**, 195302 (2003)
108. P. Cásek, S. Bouette-Russo, F. Finocchi, C. Noguera: Phys. Rev. B **69**, 085411 (2004)
109. N. Sai, B. Meyer, D. Vanderbilt, Phys. Rev. Lett. **84**, 5636 (2000)
110. R.R. Das, Y.I. Yuzyuk, P. Bhattacharya, V. Gupta, R.S. Katiyar: Phys. Rev. B **69**, 132301 (2004)
111. R. Waser, A. Rüdiger: Nature Materials **3**, 87 (2004)
112. M. Rühle, A.G. Evans: Progr. Mat. Sci. **33**, 85 (1989)
113. G. Willmann, N. Schikora, R.P. Pitto: Bioceram. **15**, 813 (1994)
114. A.M. Freborg, B.L. Ferguson, W.J. Brindley, G.J. Petrus: Mater. Sci. Eng. A **245**, 182 (1998)
115. R. Benedek, M. Minkoff, L.H. Yang: Phys. Rev. B **54**, 7697 (1996)
116. Y. Ikuhara, Y. Sugawara, I. Tanaka, P. Pirouz: Interface Science **5**, 5 (1997)
117. J.-H. Cho, K.S. Kim, C.T. Chan, Z. Zhang: Phys. Rev. B **63**, 113408 (2001)
118. C. Klein, G. Kresse, S. Surnev, F.P. Netzer, M. Schmidt, P. Varga: Phys. Rev. B **68**, 235416 (2003)
119. D.A. Muller, D.A. Shashkov, R. Benedek, L.H. Yang, J. Silcox, D.N. Seidman: Phys. Rev. Lett. **80**, 4741 (1998)
120. T. Ochs, S. Köstlmeier, C. Elsässer: Integr. Ferroelectr. **30**, 251 (2001)
121. A. Zaoui: Phys. Rev. B **69**, 115403 (2004)
122. M. Christensen, S. Dudiy, G. Wahnström: Phys. Rev. B **65**, 045408 (2002)
123. M. Christensen, G. Wahnström: Phys. Rev. B **67**, 115415 (2003)
124. J. Hartford: Phys. Rev. B **61**, 2221 (2000)
125. L. Giordano, C. DiValentin, J. Goniakowski, G. Pacchioni: Phys. Rev. Lett. **92**, 096105 (2004)
126. E. Saiz, A.P. Tomsia, R.M. Cannon: Acta Mater. **46**, 2349 (1998)
127. J.A. Venables, G.D.T. Spiller, M. Hanbucken: Rep. Prog. Phys. **47**, 399 (1984)
128. R. Schweinfest, Th. Wagner, F. Ernst: *Annual Report to the VW Foundation on the Project Progress* (Stuttgart, 1997)
129. R. Stadler, D. Vogtenhuber, R. Podloucky: Phys. Rev. B **60**, 17112 (1999)
130. R. Stadler, R. Podloucky: Phys. Rev. B **62**, 2209 (2000)
131. C. Rogero, C. Koitzsch, M.E. Gonzalez, P. Aebi, J. Cerda, J.A. Martin-Glago: Phys. Rev. B **69**, 045312 (2004)
132. C.R. Ashman, C.J. Först, K. Schwarz, P.E. Blöchl: Phys. Rev. B **69**, 075309 (2004)
133. H. Fujitani, Phys. Rev. B **57**, 8801 (1998)
134. S. Walter, F. Blobner, M. Krause, S. Muller, K. Heinz, U. Starke: J. Phys.: Condens. Matter **15**, 5207 (2003)

5

Electronic Structure and Transport for Nanoscale Device Simulation

Alex Trellakis and Peter Vogl

Summary. In this chapter, we discuss the physical models that are commonly used for the quantum simulation of electronic states and currents in nanostructures. Since most of these structures are too large for an atomistic description, we focus here on continuum models with empirically adjusted material parameters. In specific, after a short introduction into the band structure theory of crystalline solids, we first present the k·p-equations for semiconductors. Next, we discuss the envelope function approximation for heterostructures and consider the effects of elastic deformations and strain. Furthermore, we also examine carrier densities at non-zero temperature and consider the interplay of the Poisson and Schrödinger equation. After describing the electronic structure, we now discuss the Boltzmann equation and the numerically more tractable drift-diffusion equations as semi-classical models for carrier transport in semiconductors. We then extend the drift-diffusion model to take quantum corrections for size quantization into account, and we outline the principles of ballistic quantum transport. Finally, we present *nextnano*[3], a software package for the simulation of nanostructures that has been developed by the authors, and give an example application.

5.1 Introduction

Due to the rapid progress in semiconductor manufacturing technology the dimensions of electronic devices are now approaching nanometer-scale, and quantum effects are expected to have an ever increasing influence on electronic properties. For this reason, there is a lot of interest in exploring the ultimate limits of present day semiconductor technology, as exemplified by the 8 nm silicon DGFET shown in Fig. 5.1. Nanoscale MOS devices similar to the one shown are expected to form the foundations of microprocessor technology in the next decade.

But at the same time, quantum effects will eventually become dominant at small-enough length scales and prevent then conventional structures like the one shown from operating. For this reason, is also prudent to start considering completely new devices that operate solely on the basis of these

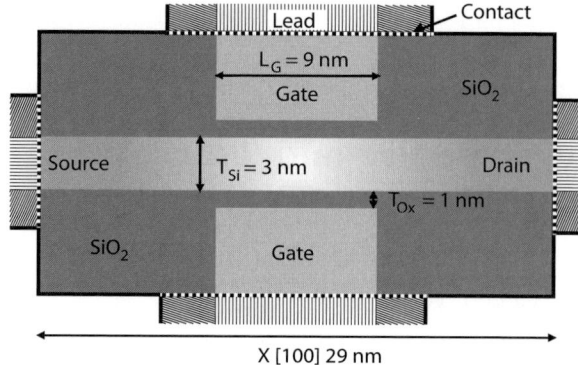

Fig. 5.1. A schematic dual gate MOSFET with a 3 nm wide channel and a 9 nm long gate

quantum effects, as for instance resonant tunneling diodes or semiconductor Mach-Zehnder interferometers [1]. Furthermore, quantum computing promises computational power that for some purposes like prime factorization surpasses the performance of today's supercomputers by many orders of magnitude.

These reasons are a major incentive for exploring the electronic properties of nanoscale semiconductor devices. Unfortunately, manufacturing a nano-device and characterizing its electronic behavior is an expensive and time-consuming undertaking in every research setting. For this reason, it is only prudent to use theoretical predictions to guide and explain the experiment. However, even nanostructures like the one shown in Fig. 5.1 contain millions of atoms. This immediately excludes the use of ab-initio methods like density functional theory. But also atomistic approaches like empirical tight binding theory or empirical pseudopotential methods become here quickly unfeasible due to their huge computational effort and their limited generality. Instead, continuum approaches like the k·p-approximation using parameters that are carefully tuned to ab-initio calculations or experiments are used for describing these mesoscopic devices.

For these reasons, state-of-the-art simulators solving continuum equations like the k·p-equations or the drift-diffusion equations have become widely used in the research community. While there are significant differences between these software packages, they all have in common that they solve a simulator-specific set of continuum equations (as for instance the multi-band Schrödinger equation, the Poisson equation, or the drift-diffusion equations) for a wide array of user-definable device structures. This allows the researcher to quickly compare the properties of different devices, or to make predictions for the outcome of experiments. And as long as one is careful enough to remain within the range of validity of the implemented continuum models (this is not always easy!), the results obtained in this fashion should compare quite well to experiment.

This overview focuses on the physical models that are used for the simulation of electronic states and currents in nanostructures. Therefore, much of the material presented here can also be found in text books for solid-state theory like [2]. In specific, part 1 starts out with a short discussion of the origin of band structures in crystalline solids and presents the k·p-equations. Here, we will also discuss a few models for the conduction and valence bands in semiconductors. In part 2, we present the envelope function approximation for heterostructures and consider the effects of elastic deformations and strain. Furthermore, we also examine carrier densities and effects of doping at non-zero temperature and consider the interplay of the Poisson and Schrödinger equation. In part 3, we present the Boltzmann equation as semi-classical transport model, and discuss the merits of drift-diffusion as a more practical alternative for larger devices. We also show how to add quantum corrections for size quantization to drift-diffusion and consider ballistic quantum transport. Finally, in part 4 we present **nextnano**[3], a software package for the simulation of nanostructures that has been developed by the authors [3]. Here, we give a short overview of the simulator's capabilities and the numerical methods that it uses, and present an example application.

5.2 Electronic Structure of Semiconductors

5.2.1 Bloch Theory and the Band Structure

Essentially all semiconductors used today are based on crystalline materials, where all constituent ions or atoms, or electronic charges are arranged in a periodic lattice. Consequently, every crystal is invariant under translations $T(\mathbf{R})$ defined by the discrete set of translation vectors

$$\mathbf{R} = n_1 \mathbf{a}_1 + n_2 \mathbf{a}_2 + n_3 \mathbf{a}_3, \qquad n_1, n_2, n_3 \in \mathbb{Z}, \tag{5.1}$$

where the three vectors $\mathbf{a}_1, \mathbf{a}_2, \mathbf{a}_3$ define the primitive cell of the crystal lattice. Furthermore, all crystal lattices are also invariant under some point operations as inversion or rotation by $60°, 90°$, or $180°$, where the exact symmetry group directly depends on the properties of the underlaying crystal lattice.

We now make the simplifying assumption that only the outermost electrons of each atom contribute significantly to the electronic properties of the crystal, while the nuclei and the inner shell electrons form a static external ion potential

$$V_{\text{ion}}(\mathbf{x} + \mathbf{R}) = V_{\text{ion}}(\mathbf{x}), \tag{5.2}$$

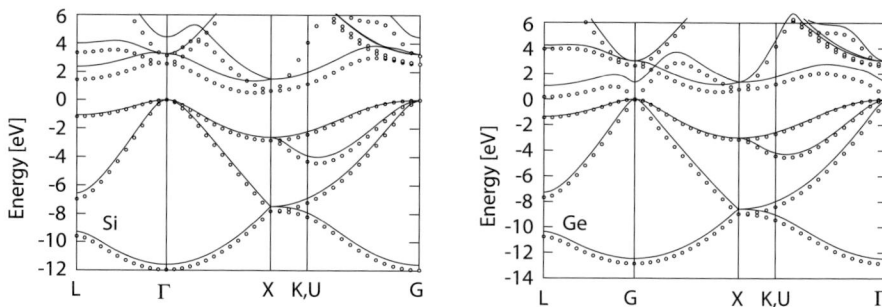

Fig. 5.2. Band structures for Si and Ge calculated with two variants of the density functional method. *Open circles* denote results for the LDA approximation and the *solid line* results from the more accurate EXX method (after [4])

that is periodic for all lattice vectors $\mathbf{R}$. In this case, the resulting many-body Hamiltonian for the outer shell electrons becomes excluding spin effects

$$\hat{H} = \sum_i \left[\frac{\hat{\mathbf{p}}_i^2}{2m} + V_{\text{ion}}(\hat{\mathbf{x}}_i) \right] + \frac{1}{2} \sum_{i \neq j} \frac{e^2}{|\hat{\mathbf{x}}_i - \hat{\mathbf{x}}_j|} \,, \tag{5.3}$$

where the last term describes the Coulomb repulsion between the electrons.

Unfortunately, the resulting N-fermion Schrödinger equation is much too complex for an exact solution without additional simplifications. For instance, in density functional theory one replaces $\hat{H}$ by a carefully constructed one-body Hamiltonian

$$\hat{H}_1 = \frac{\hat{\mathbf{p}}^2}{2m} + V_{\text{ion}}(\hat{\mathbf{x}}) + V[n](\hat{\mathbf{x}}) \,, \tag{5.4}$$

with an additional one-body potential $V[n](\mathbf{x})$. This extra potential depends self-consistently on the ground-state electron density of $\hat{H}_1$

$$n(\mathbf{x}) = \sum_{E_n < E_F} |\Psi_n(\mathbf{x})|^2 \,, \tag{5.5}$$

where the sum runs over all occupied energy levels E_n below the Fermi E_F.

One can show now that also $V[n](\hat{\mathbf{x}})$ is periodic with respect to all lattice vectors $\mathbf{R}$. Consequently, due to Bloch's theorem all eigenstates of $\hat{H}_1$ must have the form

$$\Psi_{n,\mathbf{k}}(\mathbf{x}) = e^{i\mathbf{k}\cdot\mathbf{x}} u_{n,\mathbf{k}}(\mathbf{x}) \,, \tag{5.6}$$

with

$$u_{n,\mathbf{k}}(\mathbf{x} + \mathbf{R}) = u_{n,\mathbf{k}}(\mathbf{x}) \,. \tag{5.7}$$

Furthermore, we also get that the Bloch factors $u_{n,\mathbf{k}}(\mathbf{x})$ obey

$$E_n(\mathbf{k}) u_{n,\mathbf{k}}(\mathbf{x}) = \hat{H}(\mathbf{k}) u_{n,\mathbf{k}}(\mathbf{x}) \tag{5.8}$$

$$= \left[\frac{(\hat{\mathbf{p}} + \hbar\mathbf{k})^2}{2m_0} + \hat{V}_{\text{ion}} + \hat{V}[n] \right] u_{n,\mathbf{k}}(\mathbf{x}) \,, \tag{5.9}$$

where the energy surfaces in reciprocal space $E_n(\mathbf{k})$ are also called the band structure of the semiconductor (Fig. 5.2).

5.2.2 The k·p-Approximation

However, most properties of semiconductors that are experimentally accessible depend only the position and shape of the minima and maxima $\mathbf{k}_0$ of the conduction and valence bands, and a precise knowledge of the entire band structure is not needed. For this reason, we now expand $E_n(\mathbf{k})$ around these extrema $\mathbf{k}_0$ and rewrite the Hamiltonian as

$$\hat{H}(\mathbf{k}) = \hat{H}(\mathbf{k}_0) + \frac{\hbar}{m_0}(\mathbf{k} - \mathbf{k}_0) \cdot \hat{\mathbf{p}} + \frac{\hbar^2}{2m_0}(\mathbf{k} - \mathbf{k}_0)^2 \,, \tag{5.10}$$

with eigenfunctions

$$\hat{H}(\mathbf{k}_0) u_{n,\mathbf{k}_0}(\mathbf{x}) \equiv \left[\frac{(\hat{\mathbf{p}} + \hbar\mathbf{k}_0)^2}{2m_0} + \hat{V}_{\text{ion}} + \hat{V}[n] \right] u_{n,\mathbf{k}_0}(\mathbf{x}) \tag{5.11}$$

$$= E_n(\mathbf{k}_0) u_{n,\mathbf{k}_0}(\mathbf{x}) \,. \tag{5.12}$$

Next, we use that the Bloch factors $u_{n,\mathbf{k}}(\mathbf{x})$ form a complete set of basis functions for every $\mathbf{k}$ and in specific also for $\mathbf{k}_0$. Therefore, we may insert the expansion

$$u_{n,\mathbf{k}}(\mathbf{x}) = \sum_\alpha c_{n,\alpha}(\mathbf{k}) u_{\alpha,\mathbf{k}_0}(\mathbf{x}) \tag{5.13}$$

into Schrödinger's equation (5.8) and get after a short calculation the set of equations

$$E_n(\mathbf{k}) c_{n,\alpha}(\mathbf{k}) = \sum_\beta H_{\alpha\beta}(\mathbf{k}) c_{n,\beta}(\mathbf{k}) \,,$$

with a Hamilton matrix

$$H_{\alpha\beta}(\mathbf{k}) = \left[E_\beta(\mathbf{k}_0) + \frac{\hbar^2}{2m_0}(\mathbf{k} - \mathbf{k}_0)^2 \right] \delta_{\alpha\beta} + \frac{\hbar}{m_0}(\mathbf{k} - \mathbf{k}_0) \cdot \mathbf{p}_{\alpha,\beta} \,, \tag{5.14}$$

that contains momentum matrix elements

$$\mathbf{p}_{\alpha\beta} = \int u^*_{\alpha,\mathbf{k}_0}(\mathbf{x}) \, \hat{\mathbf{p}} \, u_{\beta,\mathbf{k}_0}(\mathbf{x}) \, \mathrm{d}^3 x \,. \tag{5.15}$$

As the next step, we separate the energy bands into two groups A and B [5], where A contains a few bands of interest near the Fermi energy E_F and B all remaining bands. Using the so-called k·p-approximation, we now assume that the non-diagonal coupling term

$$\frac{\hbar}{m_0} (\mathbf{k} - \mathbf{k}_0) \cdot \mathbf{p}_{\alpha\beta} \qquad (5.16)$$

in (5.14) is weak between states from A and B and can be treated as a perturbation [6–9]. In this case, the Hamilton matrix from (5.14) now becomes the k·p-matrix

$$H_{\alpha\beta}(\mathbf{k}) = \left[E_\alpha(\mathbf{k}_0) + \frac{\hbar^2 (\mathbf{k} - \mathbf{k}_0)^2}{2m_0} \right] \delta_{\alpha\beta} + \frac{\hbar}{m_0} (\mathbf{k} - \mathbf{k}_0) \cdot \mathbf{p}_{\alpha\beta}$$
$$+ \sum_{\gamma \in B} \frac{\hat{H}_{\alpha\gamma}(\mathbf{k}) \hat{H}_{\gamma\beta}(\mathbf{k})}{E_\alpha(\mathbf{k}_0) - H_{\gamma\gamma}(\mathbf{k})} , \qquad (5.17)$$

where both indices α and β now belong to group A.

Note that the momentum matrix elements $\mathbf{p}_{\alpha\beta}$ in this expression can also be used as fitting parameters in order to match experimentally obtained results, as long one takes into account that many of the $\mathbf{p}_{\alpha\beta}$ vanish due to symmetry considerations.

5.2.3 Conduction and Valence Band Models

At this point we need to decide how many bands need to be placed in group A in order to get sufficiently accurate results. For instance, if we place only one band n into group A, we get the so-called effective mass approximation, and the k·p-approximation yields

$$H_{\text{eff}}(\mathbf{k}) = E_n(\mathbf{k}_0) + \frac{\hbar^2 (\mathbf{k} - \mathbf{k}_0)^2}{2m_0} + \frac{\hbar^2}{m_0^2} \sum_{n' \neq n} \frac{|\mathbf{p}_{nn'}|^2}{E_n(\mathbf{k}_0) - E_{n'}(\mathbf{k}_0)} \qquad (5.18)$$

$$\equiv E_n(\mathbf{k}_0) + \frac{\hbar^2}{2} \sum_{i,j=1}^{3} \left(\frac{1}{m^*} \right)_{ij} (k_i - k_{0,i})(k_j - k_{0,j}) , \qquad (5.19)$$

where the symmetric 3×3-matrix m^* is also called the effective mass of the electron.

While the effective mass approximation often suffices for the description of the conduction band, the valence bands are created by the p-orbitals in most semiconductors and are therefore threefold degenerate. For this reason, a k·p-model for the valence bands needs to include at least three bands in the orbital basis

$$|x_1\rangle, |x_2\rangle, |x_3\rangle , \qquad (5.20)$$

which for Zincblende materials at the Γ-point like GaAs then results into a k·p-Hamiltonian [10]

$$H_{\alpha\beta}^{3\times3}\left(\mathbf{k}\right) = E_\alpha\left(\mathbf{0}\right) + \frac{\hbar^2}{2m_0}\mathbf{k}^2 + N\mathbf{k}\mathbf{k}^T + \left(L - N\right)\mathrm{diag}\left(k_1^2, k_2^2, k_3^2\right) \quad (5.21)$$

$$+ M\,\mathrm{diag}\left(k_2^2 + k_3^2, k_1^2 + k_3^2, k_1^2 + k_2^2\right). \quad (5.22)$$

The parameters L,M,N here are also called the Dresselhaus parameters [7] and depend as m^* on the momentum matrix elements $\mathbf{p}_{\alpha\beta}$. In a similar fashion, we can include even more bands into the k·p-Hamiltonian (as for instance the conduction band) in order to achieve a yet more accurate band structure model.

But an accurate description of the valence bands also requires the inclusion of spin effects. Therefore, we modify our orbital basis for the valence bands as

$$|x_1 \uparrow\rangle, |x_2 \uparrow\rangle, |x_3 \uparrow\rangle, |x_1 \downarrow\rangle, |x_2 \downarrow\rangle, |x_3 \downarrow\rangle \quad (5.23)$$

and get a degenerate 6×6-Hamilton matrix

$$H_{6\times6}\left(\mathbf{k}\right) = \begin{pmatrix} H^{3\times3}\left(\mathbf{k}\right) & 0 \\ 0 & H^{3\times3}\left(\mathbf{k}\right) \end{pmatrix}, \quad (5.24)$$

with $H^{3\times3}\left(\mathbf{k}\right)$ being defined as above. This degeneracy is then partially lifted by the addition of a spin-orbit coupling term $H_{\mathrm{so}}^{6\times6}$, which results into the lowering of the split-off hole band shown in Fig. 5.3.

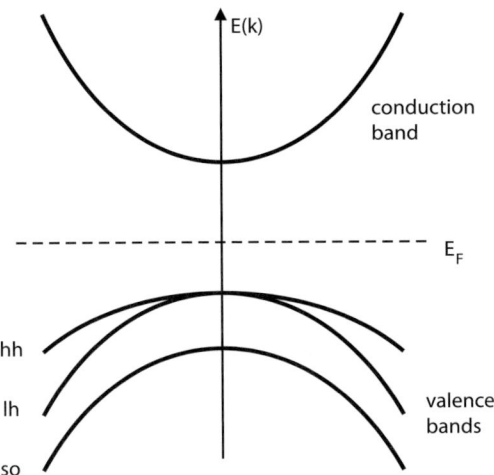

Fig. 5.3. The band structure near the Γ-point for a direct band gap semiconductor like GaAs. Shown are the conduction band, the *heavy hole* (hh), *light hole* (lh), and the *split-off hole* (so) valence bands, and the Fermi level E_F

5.3 Heterostructures

5.3.1 The Envelope Function Approximation

Until now we have only considered perfectly periodic infinite-size semiconductor crystals. However, realistic nanostructures not only have a very small finite size, but they may also contain variations in material composition and be subjected to non-uniform external potentials $V_{\text{ext}}(\mathbf{x})$. As a consequence, the Hamiltonian describing these systems is not translation invariant and neither Bloch's theorem nor equation (5.8) for the Bloch factors is available.

Fortunately, for many structures it is possible to use the envelope function approximation [6] in order to obtain results. For this approximation, we assume that a semiconductor with homogeneous material composition is subjected to an external potential $V_{\text{ext}}(\mathbf{x})$. In this case, we can write the wave function $\psi(\mathbf{x})$ as a superposition of Bloch factors with position dependent weight factors $F_\alpha(\mathbf{x})$ called envelope functions as

$$\Psi(\mathbf{x}) = \sum_{\alpha \in A} F_\alpha(\mathbf{x}) u_{\alpha, \mathbf{k}_0}(\mathbf{x}) , \qquad (5.25)$$

where the sum runs again over all bands A included in our k·p-model. If we now assume that $V_{\text{ext}}(\mathbf{x})$ varies only slowly with respect to the lattice constant, we may substitute the wave vector by the momentum operator as

$$\mathbf{k} - \mathbf{k}_0 \longrightarrow \hat{\mathbf{p}} = \frac{\hbar}{\mathrm{i}} \boldsymbol{\nabla} \qquad (5.26)$$

in our effective mass or k·p-Hamiltonian in order to obtain a real-space Hamiltonian for the envelope function $F_\alpha(\mathbf{x})$. For example, in the effective mass approximation discussed earlier we then get for the conduction band edge E_C the simple Hamiltonian

$$H_{\text{eff}}(\hat{\mathbf{p}}) = \frac{1}{2}\hat{\mathbf{p}} \cdot \left(\frac{1}{m^*}\hat{\mathbf{p}}\right) + E_\mathrm{C}(\mathbf{k}_0) + V_{\text{ext}}(\hat{\mathbf{x}}) , \qquad (5.27)$$

where we have taken into account that the effective mass m^* is actually a 3×3-matrix.

Incidentally, this approach is also widely used for the simulation of heterostructures, even though there the material composition is not constant but changes abruptly at the material interfaces (see [11, 12] for a justification). For example, in the effective mass approximation one usually makes the mass tensor m^* and the conduction band edge E_C position-dependent and uses the Hamiltonian

$$H_{\text{eff}}(\hat{\mathbf{p}}) = \frac{1}{2}\hat{\mathbf{p}} \cdot \left[\frac{1}{m^*(\hat{\mathbf{x}})}\hat{\mathbf{p}}\right] + E_\mathrm{C}(\hat{\mathbf{x}}, \mathbf{k}_0) + V_{\text{ext}}(\hat{\mathbf{x}}) \qquad (5.28)$$

in order to get quite accurate results. Similarly, we can also make all parameters in the multi-band k·p-Hamiltonians $H_{A \times A}(\hat{\mathbf{p}})$ position-dependent and obtain satisfactory results, as long as all operator ordering issues are properly taken into account.

5.3.2 Elastic Deformation and Strain

Another important physical effect that needs to be considered in the context of heterostructures are elastic deformations and therefore strain. For example, let us consider the situation where two materials with similar crystal structure but different lattice constants are epitaxially grown onto each other, as shown in Fig. 5.4. We immediately see that the crystal lattices need to compress or expand in the growth plane in order to achieve coherent growth with mutually matching chemical bonds. At the same time, we also note that every compression in the growth plane is accompanied by a corresponding expansion in the other directions and vice versa.

The elastic deformations that occur in all these cases are usually described by a displacement vector

$$\mathbf{u}\left(\mathbf{x}\right) = \mathbf{x}'\left(\mathbf{x}\right) - \mathbf{x}\,, \tag{5.29}$$

where $\mathbf{x}$ and $\mathbf{x}'$ are the positions of each mass point before and after the deformation. These deformations then results into a strain that for the typically small displacements in a semiconductor is given by the strain tensor

$$\varepsilon_{ij}\left(\mathbf{x}\right) = \frac{1}{2}\left[\partial_i u_j\left(\mathbf{x}\right) + \partial_j u_i\left(\mathbf{x}\right)\right]\,. \tag{5.30}$$

Using perturbation theory [13] one can show now that the band structure is modified in the presence of strain. In specific, we find that strain shifts the conduction band by an amount

$$E_{\mathrm{c}} \longrightarrow E_{\mathrm{c}} + a_{\mathrm{c}}\mathrm{Tr}\left(\varepsilon\right)\,, \tag{5.31}$$

where the constant a_{c} is also called the absolute deformation potential. The influence of strain on the valence band is given by an additional term

$$H^{3\times3}_{\alpha\beta}\left(\mathbf{k}\right) \longrightarrow H^{3\times3}_{\alpha\beta}\left(\mathbf{k}\right) + S_{\alpha\beta}\left(\varepsilon\right) \tag{5.32}$$

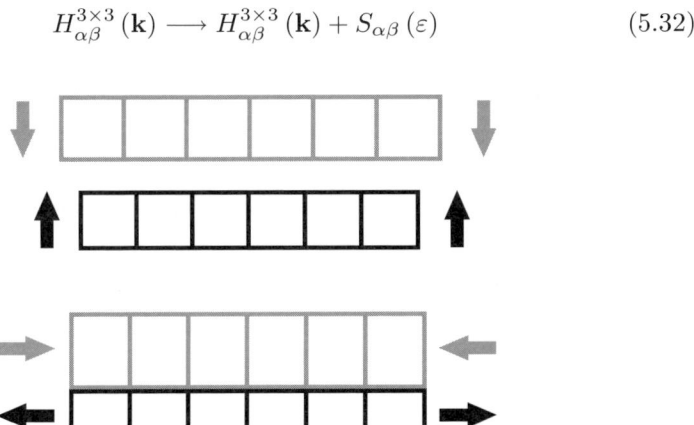

Fig. 5.4. A schematic illustration for strain effects in lattice matched heterostructures. Note that a compression in one direction is always accompanied by an expansion in the other direction and vice versa

in the k·p-matrix with

$$S_{\alpha\beta} = \begin{pmatrix} l\varepsilon_{11} + m(\varepsilon_{22} + \varepsilon_{33}) & n\varepsilon_{12} & n\varepsilon_{13} \\ n\varepsilon_{12} & l\varepsilon_{22} + m(\varepsilon_{11} + \varepsilon_{33}) & n\varepsilon_{23} \\ n\varepsilon_{13} & n\varepsilon_{23} & l\varepsilon_{33} + m(\varepsilon_{11} + \varepsilon_{22}) \end{pmatrix},$$

(5.33)

where the parameters l, m, n are connected to the absolute deformation potential of the valence band a_v and its shear deformation potentials b, d by

$$l = a_v + 2b \qquad m = a_v - b \qquad n = \sqrt{3}d.$$

(5.34)

Thus we find here that strain can lift the degeneracy between light holes and heavy holes in the valence band as shown in Fig. 5.5.

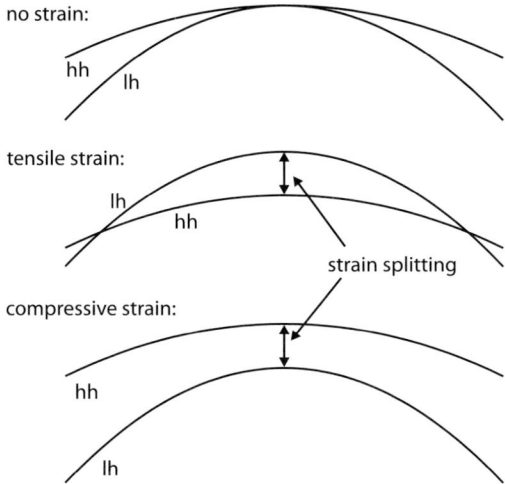

Fig. 5.5. The effects of strain on the valence bands in GaAs for different types of strain

5.3.3 Carrier Densities at Non-Zero Temperature

So far we have ignored the effects of a non-zero temperature in semiconductor structures. At zero temperature and in absence of external electromagnetic fields, the conduction band is devoid of electrons while the valence bands are completely occupied due to the Fermi energy E_F being between both band edges as

$$E_V < E_F < E_C.$$

(5.35)

Consequently, neither band can carry an electric current and the semiconductor becomes an insulator.

For non-zero temperatures this situation changes significantly. Now we find that sufficiently energetic lattice vibrations (or phonons) can lift electrons from a valence band into the conduction band, while leaving a positively charged hole in the valence band behind (see Fig. 5.6), a process that is also called (pair) generation. These excess electrons and the holes can now be easily accelerated by electromagnetic fields, which can lead to a buildup of space charges and the presence of electric currents. Finally, an electron-hole pair may also recombine under the emission of a phonon or photon, with the electron filling up the gap that originally belonged to the hole.

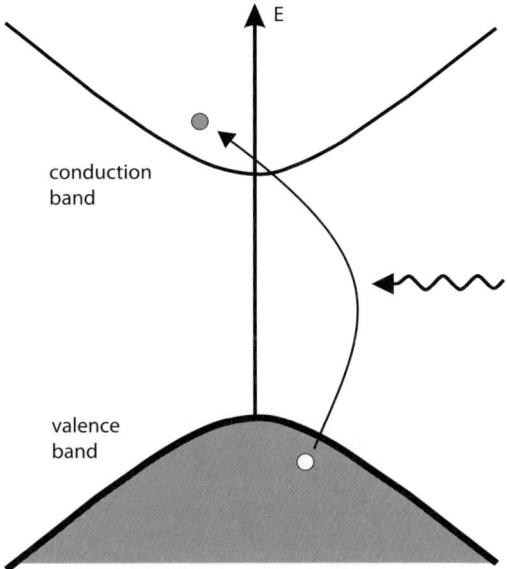

Fig. 5.6. An electron is excited from a valence band to the conduction band, leaving a hole at its original place

In thermal equilibrium, the generation and recombination processes balance each other and the occupation $N_\mathrm{e}\left(E\right)$ for a state with energy E is given by the Fermi distribution as

$$N_\mathrm{e}\left(E\right) = \frac{2}{1 + \exp\left(\frac{E - E_\mathrm{F}}{k_\mathrm{B} T}\right)}, \tag{5.36}$$

where k_B is the Boltzmann constant and T the temperature of the system. If we now use a suitable envelope function Hamiltonian $\hat{H}_\mathrm{C}$ in order to determine the wave functions of the conduction band electrons

$$\hat{H}_\mathrm{C}\psi_n^\mathrm{e}\left(\mathbf{x}\right) = E_n^\mathrm{e}\psi_n^\mathrm{e}\left(\mathbf{x}\right), \tag{5.37}$$

we then get for the corresponding electron density

$$n\left(\mathbf{x}\right) = \sum_n N_{\mathrm{e}}\left(E_n^{\mathrm{e}}\right)\left|\psi_n^{\mathrm{e}}\left(\mathbf{x}\right)\right|^2 . \qquad (5.38)$$

At the same time, the occupation $N_{\mathrm{h}}\left(E\right)$ of a hole state at energy E in a valence band is given by a Fermi distribution as well with

$$N_{\mathrm{h}}\left(E\right) = \frac{2}{1 + \exp\left(\frac{E_{\mathrm{F}} - E}{k_{\mathrm{B}}T}\right)} . \qquad (5.39)$$

Therefore, starting from a suitable k·p-Hamiltonian $\hat{H}_{\mathrm{V}}$ for the valence bands with

$$\hat{H}_{\mathrm{V}}\psi_n^{\mathrm{h}}\left(\mathbf{x}\right) = E_n^{\mathrm{h}}\psi_n^{\mathrm{h}}\left(\mathbf{x}\right) , \qquad (5.40)$$

we now get as expected

$$p\left(\mathbf{x}\right) = \sum_n N_{\mathrm{h}}\left(E_n^{\mathrm{h}}\right)\left|\psi_n^{\mathrm{h}}\left(\mathbf{x}\right)\right|^2 \qquad (5.41)$$

for the hole density.

The number of free carriers in a semiconductor is further modified by the presence of dopant atoms. Here, electron donors require a small ionization energy E_{d} to donate an electron to the conduction band, while electron acceptors need a small energy E_{a} remove an electron from the valence band and create a hole in its place. The ionization energies E_{d} and E_{a} are small compared to the band gap for both kind of species, which results in the dopant atoms to be mostly ionized at room temperature. As a result, electron donors lead to a massive increase of the electron density (and a corresponding decrease in the hole density), while electron acceptors result in a large decrease of the hole density (and a corresponding decrease in the electron density). Finally, as a side effect of their almost complete ionization at room temperature dopants also result in a positive space charge $N_{\mathrm{D}}^+\left(\mathbf{x}\right)$ in the case of donors and a negative space charge $N_{\mathrm{A}}^-\left(\mathbf{x}\right)$ in the case of acceptors.

5.3.4 Charge Distributions and the Poisson Equation

All these space charges together with the free electrons and holes and possible surface charges $\rho_{\mathrm{s}}\left(\mathbf{x}\right)$ add up to a total charge

$$\rho\left(\mathbf{x}\right) = e\left[p\left(\mathbf{x}\right) - n\left(\mathbf{x}\right) + N_{\mathrm{D}}^+\left(\mathbf{x}\right) - N_{\mathrm{A}}^-\left(\mathbf{x}\right) + \rho_{\mathrm{s}}\left(\mathbf{x}\right)\right] , \qquad (5.42)$$

that can be quite large at some places. As a result, we get according to Poisson's equation

$$\boldsymbol{\nabla}\cdot\left[\epsilon\left(\mathbf{x}\right)\boldsymbol{\nabla}\phi\left(\mathbf{x}\right)\right] = -\rho\left(\mathbf{x}\right) \qquad (5.43)$$

an electric potential $\phi(\mathbf{x})$ that warps the conduction and valence band edges E_C and E_V as

$$E_C \longrightarrow E_C[\phi] = E_C - e\phi \tag{5.44}$$

$$E_V \longrightarrow E_V[\phi] = E_V - e\phi. \tag{5.45}$$

As consequence, both the carrier densities and the ionized dopant space charges become potential-dependent, and the Poisson equation (5.43) now becomes nonlinear in the electrostatic potential ϕ as

$$\mathbf{\nabla} \cdot (\epsilon \mathbf{\nabla} \phi) = -e \left(p[\phi] - n[\phi] + N_D^+[\phi] - N_A^-[\phi] + \rho_s \right). \tag{5.46}$$

Here, the dependence of the carrier densities n and p on ϕ is a consequence of $E_C[\phi]$ and $E_V[\phi]$ entering the Hamiltonian in Schrödinger's equation as

$$\hat{H}_C \longrightarrow \hat{H}_C[\phi] = \hat{H}_C - e\hat{\phi} \tag{5.47}$$

$$\hat{H}_V \longrightarrow \hat{H}_V[\phi] = \hat{H}_V - e\hat{\phi}, \tag{5.48}$$

which then of course makes also the electron and hole wave functions and their energies ϕ-dependent as

$$\psi_n^e = \psi_n^e[\phi] \qquad E_n^e = E_n^e[\phi] \tag{5.49}$$

$$\psi_n^h = \psi_n^h[\phi] \qquad E_n^h = E_n^h[\phi]. \tag{5.50}$$

Therefore, we not only need to solve a nonlinear Poisson equation (5.46) for ϕ, we also need to solve (5.46) together with the Schrödinger equation in order to obtain a correct self-consistent set of quantum states and carrier densities.

5.4 Carrier Transport in Nanostructures

5.4.1 Classical Ballistic Transport

The reader might have noticed that in the last section about heterostructures we have assumed thermal equilibrium with a position independent temperature T and Fermi level E_F. But of course all semiconductor devices like the one shown in Fig. 5.1 have at least two contacts where an external bias can be applied. If the applied bias is not the same for all contacts, we immediately get a non-equilibrium situation where electrical currents flow and the carrier distributions may become time-dependent.

Developing an accurate quantum transport theory for these systems now requires the additional inclusion of phonons to the many-body Schrödinger equation describing the physical system. A model that can been employed for this type of system is for instance the non-equilibrium Green's function approach (see [14] and the text books [15, 16] for in-depth expositions).

But we can already achieve a first understanding of carrier transport by starting out with a classical ballistic transport model, where we assume that all carriers move as compact uncertainty-limited wave packets along classical orbits $\mathbf{x}(t)$ and $\mathbf{k}(t)$ in phase space. These classical orbits obey for a single energy band $E(\mathbf{k})$ the equations

$$\dot{\mathbf{x}} = \frac{1}{\hbar} \nabla_{\mathbf{k}} E(\mathbf{k}) \qquad (5.51)$$

$$\hbar \dot{\mathbf{k}} = q\mathbf{E}(\mathbf{x}, t) \,, \qquad (5.52)$$

where the second equation is the electrostatic force on a particle with charge q.

As we have discussed earlier, $E(\mathbf{k})$ is periodic in reciprocal space but can be approximated near a minimum $\mathbf{k}_0$ by

$$E(\mathbf{k}) \approx \frac{\hbar^2}{2}(\mathbf{k} - \mathbf{k}_0) \cdot \left[\frac{1}{m^*}(\mathbf{k} - \mathbf{k}_0)\right] \,, \qquad (5.53)$$

which then leads to a simplified equation of motion

$$\dot{\mathbf{x}} = \frac{\hbar}{m^*}(\mathbf{k} - \mathbf{k}_0) \qquad (5.54)$$

or

$$m^* \ddot{\mathbf{x}} = q\mathbf{E}(\mathbf{r}, t) \,, \qquad (5.55)$$

which is the Newton force law for a anisotropic mass tensor m^*.

5.4.2 Scattering and the Boltzmann Equation

We now add carrier-phonon scattering as discrete events in time where a carrier wave packet undergoes an instantaneous transition from a reciprocal space point $\mathbf{k}$ to another $\mathbf{k}'$. The transition probability $S(\mathbf{k}, \mathbf{k}')$ for these scattering events derives of course from quantum mechanics and is accompanied by a phase change $\delta\varphi$ which we assume to be random for simplicity.

The inelastic mean free path between these scattering events (and therefore also the coherence length) now varies for different semiconductors and scattering mechanisms typically between about 3 and 100 nm which is much larger than a single lattice constant. This justifies the use of a band structure $E(\mathbf{k})$ between scattering events.

At the same time, the size of the wave packet in phase space needs to obey the uncertainty relation

$$\Delta x \Delta k \approx 1 \,, \qquad (5.56)$$

where the upper limit for Δx is given by the coherence length and the one for Δk by the Brillouin zone boundary at

$$k_{\mathrm{max}} \approx 10^{10} \ \mathrm{m}^{-1} \,. \qquad (5.57)$$

If we now choose Δx matching to the coherence length as

$$\Delta x \approx 10 - 100 \text{ nm} , \tag{5.58}$$

we get from (5.56)

$$\Delta k \approx 10^7 - 10^8 \text{ m}^{-1} \ll k_{\max} , \tag{5.59}$$

which is sufficiently accurate to sample reciprocal space. Therefore, as long as we do not attempt to localize the particle in phase space with higher accuracy as given by the pair Δx and Δk, we can define a distribution function $f(\mathbf{x}, \mathbf{k}, t)$ that represents the probability to find the particle between the phase space points $(\mathbf{x}, \mathbf{k})$ and $(\mathbf{x} + d\mathbf{x}, \mathbf{k} + d\mathbf{k})$.

Flux conservation in phase space due to Liouville's theorem gives us now immediately the Boltzmann equation

$$\frac{\mathrm{d}f}{\mathrm{d}t} \equiv \frac{\partial f}{\partial t} + \dot{\mathbf{x}} \cdot \boldsymbol{\nabla}_{\mathbf{x}} f + \dot{\mathbf{k}} \cdot \boldsymbol{\nabla}_{\mathbf{k}} f \tag{5.60}$$

$$= \frac{\partial f}{\partial t} + \frac{1}{\hbar} \boldsymbol{\nabla}_{\mathbf{k}} E(\mathbf{k}) \cdot \boldsymbol{\nabla}_{\mathbf{x}} f + \frac{q}{\hbar} \mathbf{E}(\mathbf{x}, t) \cdot \boldsymbol{\nabla}_{\mathbf{k}} f = \left. \frac{\mathrm{d}f}{\mathrm{d}t} \right|_{\text{scatt}} , \tag{5.61}$$

where we have collected all variations in f due to scattering in the term at the right. For this scattering term one can easily show that it must be equal to

$$\left. \frac{\mathrm{d}f(\mathbf{x}, \mathbf{k}, t)}{\mathrm{d}t} \right|_{\text{scatt}} = \int \mathrm{d}^3 k' \, f(\mathbf{x}, \mathbf{k}', t) \, S(\mathbf{k}', \mathbf{k}) \left[1 - f(\mathbf{x}, \mathbf{k}, t) \right] \tag{5.62}$$

$$- \int \mathrm{d}^3 k' f(\mathbf{x}, \mathbf{k}, t) \, S(\mathbf{k}, \mathbf{k}') \left[1 - f(\mathbf{x}, \mathbf{k}', t) \right] , \tag{5.63}$$

where the integrals run over the entire first Brillouin zone, and the factors $1 - f$ are a direct consequence of the Pauli exclusion principle.

The Boltzmann equation as presented in (5.61) is a time-dependent six-dimensional integro-differential equation that due to its complexity cannot be solved accurately (even though approximate solutions obtained using Monte Carlo methods suffice for some applications). This situation is aggravated by the fact that the (now classical) electron density

$$n(\mathbf{x}) = \int \mathrm{d}^3 k \, f(\mathbf{x}, \mathbf{k}, t) \tag{5.64}$$

is a source term in Poisson's equation (5.43). Since the gradient of the electrostatic potential ϕ enters the Boltzmann equation (5.61) via $\mathbf{E} = -\boldsymbol{\nabla}\phi$, a self-consistent solution of both partial differential equations is required.

In addition, if both electrons and holes are present we need two phase space densities f_{n} and f_{p} that each obey their own Boltzmann equation with different band structures $E_{\mathrm{n}}(\mathbf{k})$ and $E_{\mathrm{p}}(\mathbf{k})$. These two equations couple then directly through pair generation and recombination and indirectly through $\boldsymbol{\nabla}\phi$ from Poisson's equation (5.43).

Finally, is important to emphasize that the size of most nanostructures is comparable or smaller than the $10 - 100$ nm we have assumed earlier for Δx. Therefore, some quantum corrections to the Boltzmann equation (5.61) are needed in order to obtain reasonable results.

5.4.3 The Drift-Diffusion Equations

For semiconductor devices that are large enough compared to Δx we can often replace Boltzmann's equation by the much simpler drift-diffusion equations. For this drift-diffusion model, we assume that the electric fields $\mathbf{E}$ are small enough and the scattering events frequent enough that the system is locally at each point $\mathbf{x}$ almost in thermal equilibrium. With other words, we assume that the phase space distributions f_n and f_p are approximately Fermi distributed in $\mathbf{k}$-space for each $\mathbf{x}$, with quasi-Fermi levels $E_{F,n}(\mathbf{x}, t)$ and $E_{F,p}(\mathbf{x}, t)$ that are now position-dependent. In this case, we get for the electron and hole density in quasi-equilibrium

$$
n(\mathbf{x}, t) = \int_0^\infty dE \frac{g_C(E)}{1 + \exp\left[\frac{E_C(\mathbf{x},t) - E_{F,n}(\mathbf{x},t) + E}{k_B T}\right]} \tag{5.65}
$$

$$
p(\mathbf{x}, t) = \int_0^\infty dE \frac{g_V(E)}{1 + \exp\left[\frac{E_{F,p}(\mathbf{x},t) - E_V(\mathbf{x},t) + E}{k_B T}\right]}, \tag{5.66}
$$

where $E_C(\mathbf{x})$ and $E_V(\mathbf{x})$ are the conduction and valence band edges, and $g_C(E)$ and $g_V(E)$ the density-of-states for electrons and holes.

Furthermore, since the electron and hole phase space densities f_n and f_p deviate only slightly from the Fermi distribution in $\mathbf{k}$-space, we can make a moment expansion which then yields in zeroth order the following continuity equations for electrons and holes

$$
\frac{\partial n(\mathbf{x}, t)}{\partial t} + \boldsymbol{\nabla} \cdot \mathbf{j}_n(\mathbf{x}, t) = G(\mathbf{x}, t) - R(\mathbf{x}, t) \tag{5.67}
$$

$$
\frac{\partial p(\mathbf{x}, t)}{\partial t} + \boldsymbol{\nabla} \cdot \mathbf{j}_p(\mathbf{x}, t) = G(\mathbf{x}, t) - R(\mathbf{x}, t), \tag{5.68}
$$

where $\mathbf{j}_n$ and $\mathbf{j}_p$ are the electrons and hole particle currents and G and R the pair generation and recombination rates. The currents themselves are given by the first order component of the moment expansion as

$$
\mathbf{j}_n(\mathbf{x}, t) = -\mu_n(\mathbf{x}) n(\mathbf{x}, t) \boldsymbol{\nabla} E_{F,n}(\mathbf{x}, t) \tag{5.69}
$$

$$
\mathbf{j}_p(\mathbf{x}, t) = \mu_p(\mathbf{x}) p(\mathbf{x}, t) \boldsymbol{\nabla} E_{F,p}(\mathbf{x}, t), \tag{5.70}
$$

where $\mu_n(\mathbf{x})$ and $\mu_p(\mathbf{x})$ are the electron and hole mobilities.

This set of equations (5.67–5.70) are called the drift-diffusion equations, and form together with the Poisson equation (5.43) the basis of many commercially available semiconductor device simulators. The accuracy of the drift-diffusion approximation can be further improved by using complicated phenomenological models for the carrier mobilities and generation and recombination rates. Furthermore, non-uniform carrier temperatures and heat transport can also be included if needed. Especially the commercially available simulators offer here many options for silicon based MOS transistors.

5.4.4 Quantum Corrected Drift-Diffusion

But as we have seen earlier, for nanoscale semiconductor structures quantum mechanics will lead to major deviations from the Boltzmann equation and therefore also from the drift-diffusion equations. Luckily enough, there are important mesoscopic structures like the DGFET shown in Fig. 5.1 where quantum mechanical size quantization must be included but no ballistic transport or resonant tunneling takes place. For such physical systems, a quantum correction to the *stationary* drift-diffusion equations is still possible and a full quantum transport calculation can be avoided [17, 18].

The key to finding this quantum correction lies in the realization that the electron and hole densities in (5.67–5.70) are purely classical and can therefore not take into account size quantization. We can correct this deficiency by using the electron and hole Hamiltonians $\hat{H}_C[\phi]$ and $\hat{H}_V[\phi]$ for calculating the electron and hole envelope wave functions as

$$\hat{H}_C[\phi]\,\psi_n^e(\mathbf{x}) = E_n^e\psi_n^e(\mathbf{x}) \tag{5.71}$$

$$\hat{H}_V[\phi]\,\psi_n^h(\mathbf{x}) = E_n^h\psi_n^h(\mathbf{x}) \tag{5.72}$$

and then determining the corresponding carrier densities through the electron and hole Fermi levels $E_{F,n}(\mathbf{x})$ and $E_{F,p}(\mathbf{x})$ using formulas (5.36–5.41) as

$$n(\mathbf{x}) = 2\sum_n \frac{|\psi_n^e(\mathbf{x})|^2}{1 + \exp\left(\frac{E_n^e - E_{F,n}(\mathbf{x})}{k_B T}\right)} \tag{5.73}$$

and

$$p(\mathbf{x}) = 2\sum_n \frac{|\psi_n^h(\mathbf{x})|^2}{1 + \exp\left(\frac{E_{F,p}(\mathbf{x}) - E_n^h}{k_B T}\right)}, \tag{5.74}$$

where we have implicitly also used the adiabatic approximation for the wave function.

These quantum densities can now be entered into the drift-diffusion equations (5.67–5.70) which for stationary currents become with

$$\boldsymbol{\nabla} \cdot \mathbf{j}_n(\mathbf{x}) = -\boldsymbol{\nabla} \cdot [\mu_n(\mathbf{x})\,n(\mathbf{x})\,\boldsymbol{\nabla}E_{F,n}(\mathbf{x})] \equiv G(\mathbf{x}) - R(\mathbf{x}) \tag{5.75}$$

and

$$\nabla \cdot \mathbf{j}_{\mathrm{p}}(\mathbf{x}) = \nabla \cdot [\mu_{\mathrm{p}}(\mathbf{x}) \, p(\mathbf{x}) \, \nabla E_{\mathrm{F,p}}(\mathbf{x})] \equiv G(\mathbf{x}) - R(\mathbf{x}) \qquad (5.76)$$

two boundary value problems for the Fermi levels $E_{\mathrm{F,n}}(\mathbf{x})$ and $E_{\mathrm{F,p}}(\mathbf{x})$. These two equations now need to be solved self-consistently with the nonlinear Poisson equation (5.46)

$$\nabla \cdot (\epsilon \nabla \phi) = -e \left(p - n + N_{\mathrm{D}}^{+}[\phi] - N_{\mathrm{A}}^{-}[\phi] + \rho_{\mathrm{s}}\right) \qquad (5.77)$$

in order to obtain the final quantum-corrected carrier densities and currents. See for example [19] for simulation results that have obtained in this fashion for a double-gate MOSFET similar to the one shown in Fig. 5.1.

5.4.5 Quantum Ballistic Transport

Finally, a full quantum transport calculation can also be avoided if the semiconductor structure is small enough or the temperature low enough that no phonon scattering occurs and transport is fully ballistic in the absence of impurities. This situation occurs for instance in quantum interference devices [1] or also in quantum resonant tunneling structures [20] like the one shown in Fig. 5.7.

For these systems, we can use the Landau–Büttiker formalism [21, 22] to calculate the ballistic current $J_{\lambda\lambda'}$ between all contacts λ and λ' with the Landauer formula. This formula generally involves an integration over all quantum numbers that characterize the lead states and is given its most general form in [23]. Showing only the energy dependence in the integral for simplicity, we get here for the ballistic current

$$J_{\lambda\lambda'} = \frac{e}{\pi\hbar} \int \mathrm{d}E \, T_{\lambda\lambda'}(E) \left[f_{\lambda}(E) - f_{\lambda'}(E)\right], \qquad (5.78)$$

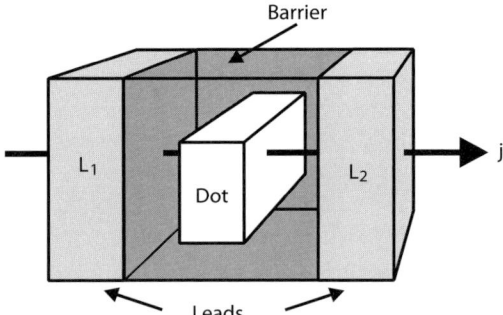

Fig. 5.7. A model quantum dot resonant tunneling structure. The quantum dot has the dimensions $14 \times 14 \times 5$ nm and is embedded in a 15 nm thick barrier region (after [20])

where E is the energy, $f_\lambda(E)$ the equilibrium distribution function inside contact λ, and $T_{\lambda\lambda'}(E)$ the transmission function between contact λ and λ'.

These transmission functions are then defined using the retarded single particle Green's function (see again [15])

$$\hat{G}^R(E) = \frac{1}{E - \hat{H} + \mathrm{i}\eta} \qquad (5.79)$$

as

$$T_{\lambda\lambda'} = \mathrm{Tr}\left(\hat{\Gamma}_\lambda \hat{G}^R \hat{\Gamma}_{\lambda'} \hat{G}^{R\dagger}\right), \qquad \lambda \neq \lambda', \qquad (5.80)$$

where the operators $\hat{\Gamma}_\lambda$ are related to the self-energy matrices $\hat{\Sigma}_\lambda$ in the contacts as

$$\hat{\Gamma}_\lambda = \mathrm{i}\left(\hat{\Sigma}_\lambda - \hat{\Sigma}_\lambda^\dagger\right). \qquad (5.81)$$

Since we have ballistic transport both inside of the device and also in the contact regions, scattering only occurs at the contact-device boundaries and $\hat{\Sigma}_\lambda$ disappears everywhere else. This then allows us to use the contact block reduction method [24] for determining $T_{\lambda\lambda'}$ (see Fig. 5.8).

5.5 The nextnano^3 Simulation Package

5.5.1 Capabilities Overview

Most of the physical models that we have discussed so far have been implemented into **nextnano^3**, a software package that the authors have specifically designed for the simulation of realistic nanoscale heterostructures in one, two,

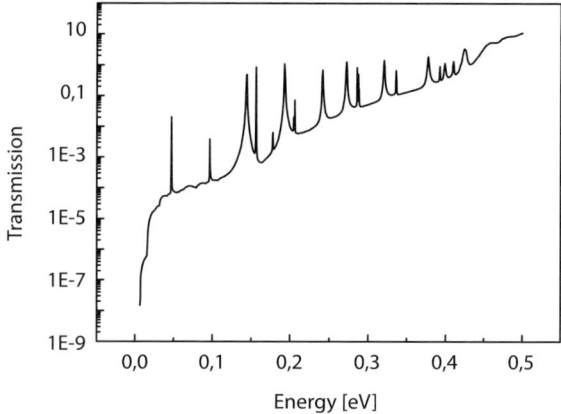

Fig. 5.8. The transmission function for the resonant tunneling structure shown in Fig. 5.7 (after [20])

and three dimensions [3, 18]. In specific, the electronic structure is calculated by solving self-consistently the coupled system of the nonlinear Poisson equation (5.77) and Schrödinger's equation. For the Schrödinger's equation, we use either the effective mass approximation for the electrons (5.18) and the 6-band k·p-approximation for the holes (5.22), or an 8-band k·p-approximation for both electrons and holes at the same time.

In addition, we also perform strain relaxation in order to determine the strain tensor $\varepsilon_{ij}\,(\mathbf{x})$ and the strain-dependent changes in the conduction (5.31) and valence band (5.33). The presence of strain-induced piezoelectric charges and the effects of the spin-orbit interaction on the valence bands can be taken into account as well.

Furthermore, the quantum-corrected drift-diffusion equations (5.75–5.76) described above have been implemented to allow the determination of quasi-Fermi levels and particle currents for large classical devices and also many mesoscopic quantum structures. In addition, various phenomenological models for mobility, generation, and recombination are available in order to obtain more accurate simulation results.

The parameters entering these models and describing the material properties of various semiconductors and semiconductor alloys are automatically read in from a large materials database that can also be modified and extended by the software user. Similarly, a description of the device geometry, doping and alloy profiles, and simulation options is read from an external input file that can be easily modified by the user as well.

5.5.2 Numerical Methods

The numerical methods used in **nextnano**[3] follow from the simple fact that the equations we need to solve are sets of coupled partial differential equations (PDEs) in position space. This immediately necessitates their discretization on a grid. In order to achieve a highly efficient but also simple software implementation only nonuniform tensor grids are used. The mapping of the PDEs onto the grid is achieved using box discretization, since this discretization scheme is flux conserving in the presence of material discontinuities as they naturally occur in the simulation of heterostructures.

As a result of this discretization procedure, every differential operator in these PDEs now becomes a sparse $N \times N$-matrix, where N is the number of nodes in the grid. This number of grid nodes now scales as

$$N \sim \frac{1}{(\Delta x)^D}, \tag{5.82}$$

where Δx is the desired space accuracy and D the number of space dimensions. Here we see immediately that three-dimensional simulations require extremely large matrices that cannot be easily stored and processed, unlike one explicitly exploits the sparsity structure of the matrices with its many vanishing entries. Similarly, every rank-r function over real space becomes after discretization an rN-dimensional vector, with $r = 1$ for scalar function like the electrostatic potential or the quasi-Fermi levels, $r = D$ for vector-valued functions like the particle currents, and $r = D^2$ for rank-2 tensor fields like the strain tensor.

Therefore, every PDE that is a boundary value problem (like the Poisson or the stationary drift-diffusion equation) becomes after discretization a large system of linear or nonlinear equations (N equations and unknowns), while the Schrödinger equation as an operator eigenvalue problem becomes a similarly large matrix eigenvalue problem with discretized wave functions as eigenvectors. Due to the sparsity of the underlying system matrices, these large matrix problems can now be efficiently solved using specialized linear algebra routines for sparse matrices like the conjugate gradient method or the Arnoldi iteration (see [25] for an in-depth presentation).

In addition, standard algorithms like the Newton method with line search are used to deal with nonlinearity like in Poisson's equation, while elaborate iteration schemes like the predictor-corrector approach [26] are employed in order to avoid numerical instabilities in finding a self-consistent solution. As recapitulation, an overview of the program flow in **nextnano**[3] is also shown in Fig. 5.9.

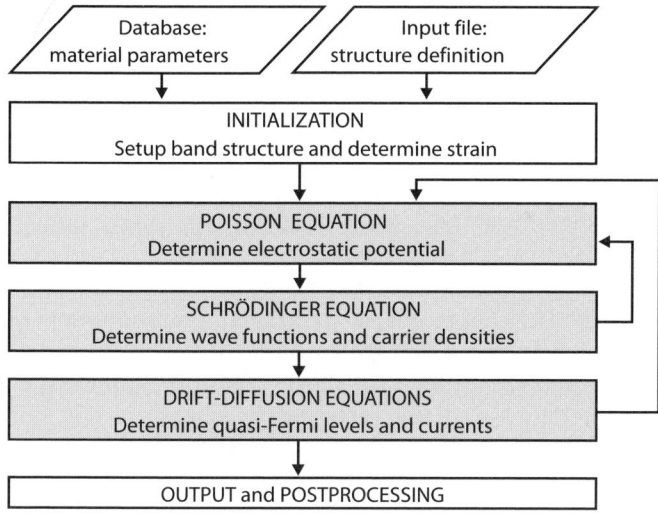

Fig. 5.9. An overview of the program flow in **nextnano**[3]

5.5.3 Example Application

As example application for the use of **nextnano**[3], we simulate the mechanical and electronic properties of self-assembled quantum dots. Optoelectronic devices bases on quantum dots have already demonstrated superior characteristics compared to conventional systems that give rise to the expectation that quantum dot lasers and modulators will be widely used in the future. Furthermore, quantum dots are also supposed to be the main elements in single-electron systems needed for quantum computing.

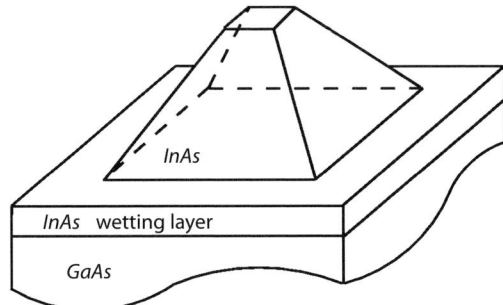

Fig. 5.10. A schematic plot of a self-assembled quantum dot

For this reason, there is a lot of interest in the study of quantum dots, as for instance self-assembled dots as the one shown in Fig. 5.10. This specific type of dot is high strained due to the lattice mismatch between InAs dot and GaAs substrate which gives rise to a large piezoelectric charge. Since these piezoelectric charges also modify the electrostatic potential that is felt by electrons or holes bound in the dot, we can expect that the shape of the electron and hole wave function is strongly influenced by the strain.

Using **nextnano**[3], M. Povolotskyi et al. [27] have calculated the strain and the piezoelectric charge for a pyramidal quantum dot as shown in Fig. 5.10. Their results for a dot grown on a (311)-GaAs substrate is shown in Figs. 5.11 and 5.12. Here it is most noteworthy that the strain distribution and therefore also the piezoelectric charge is not symmetric despite the symmetric dot shape. As an immediate result, the electrostatic potential and therefore eventually also the electron and hole wave functions will not be symmetric. This asymmetry can have a huge influence on the overlap of electron and hole wave functions and therefore also on the dot's exciton absorption, which is crucial for the dot's optoelectronic performance.

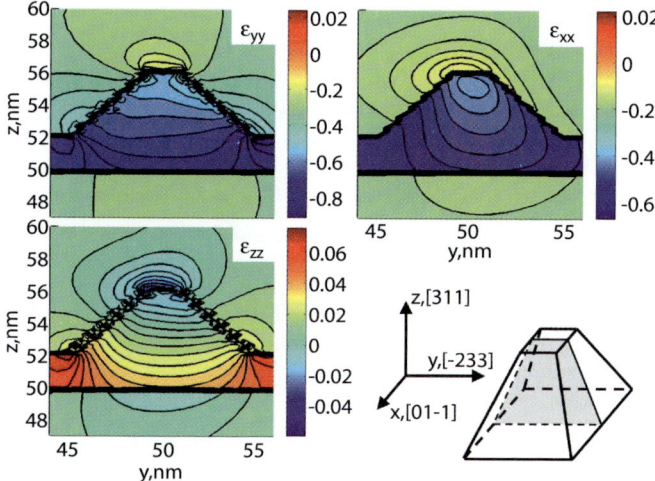

Fig. 5.11. Components of the strain tensor for a quantum dot with a composition and geometry as shown in Fig. 5.10. The dot is assumed to have grown on a (311)-substrate (after [27])

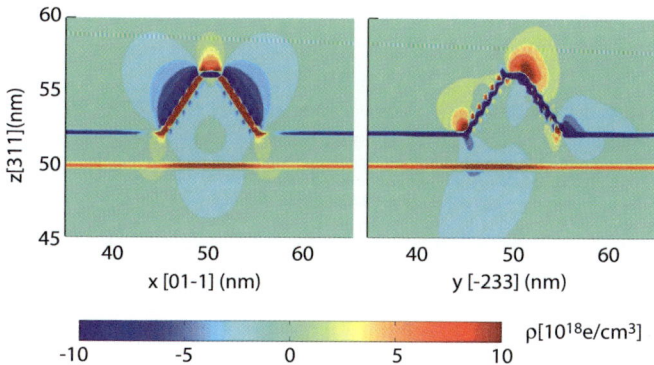

Fig. 5.12. The piezoelectric charge that is induced by the strain shown in Fig. 5.11 (after [27])

References

1. M. Sabathil, D. Mamaluy, P. Vogl: Semicond. Sci. Technol. **19**, S137 (2004)
2. P.Y. Yu, M. Cardona: Fundamentals of Semiconductors (Springer, Berlin Heidelberg New York 1996)
3. The **nextnano**^{3} simulation package and example input files can be downloaded from http://www.wsi.tum.de/nextnano3/
4. J.A. Majewski, A. Trellakis, P. Vogl et al.: physica status solidi (c) **1**, 2003 (2004)

5. P. Löwdin: J. Chem. Phys. **19**, 1396 (1951)
6. J.M. Luttinger, W. Kohn: Phys. Rev. **97**, 869 (1955)
7. G. Dresselhaus, A.F. Kip, C. Kittel: Phys. Rev. **98**, 368 (1955)
8. C.R. Pidgeon, R.N. Brown: Phys. Rev. **146**, 575 (1966)
9. G. Bastard: Phys. Rev. B **24**, 5693 (1981)
10. E.O. Kane: The k·p method. In: *Semiconductors and Semimetals: III-V compounds,* vol. 1, ed by R.K. Willardson, A.C. Beer (Academic Press, New York 1966) pp. 75-100
11. M.G. Burt: J. Phys.: Condens. Matter **4**, 6651 (1992)
12. M.G. Burt: J. Phys.: Condens. Matter **11**, R53 (1999)
13. T.B. Bahder: Phys. Rev. B **41**, 11992 (1990)
14. R. Lake et al: J. Appl. Phys. **81**, 7845 (1997)
15. S. Datta: *Electronic Transport in Mesoscopic Systems,* (Cambridge University Press, Cambridge 1997)
16. D.K. Ferry, S.M. Goodnick: *Transport in Nanostructures,* (Cambridge University Press, Cambridge 2000)
17. S. Hackenbuchner: Elektronische Struktur von Halbleiter-Nanobauelementen im thermodynamischen Nichtgleichgewicht. Ph.D. Thesis, TU München, München (2002)
18. M. Sabathil, S. Hackenbuchner, P. Vogl et al.: J. Comp. Electronics **1**, 81 (2002)
19. K. Kim, O. Kwon, A. Trellakis et al.: J. Korean Phys. Soc. **45**, S909 (2004)
20. M. Sabathil, S. Birner, P. Vogl et al.: J. Comp. Electronics **2**, 269 (2003)
21. R. Landauer: Phys. Scr., T **42**, 110 (1992)
22. M. Büttiker: IBM J. Res. Dev. **32**, 317 (1988)
23. A. Di Carlo, P. Vogl: Phys. Rev. B **50**, 8358 (1994)
24. D. Mamaluy, M. Sabathil, P. Vogl: Phys. Rev. B **93**, 4628 (2003)
25. L.N. Trefethen, D. Bau III: *Numerical Linear Algebra,* (SIAM Society for Industrial & Applied Mathematics, Philadelphia 1997)
26. A. Trellakis et al: J. Appl. Phys. **81** 7880 (1997)
27. M. Povolotskyi, A. Di Carlo, P. Vogl et al.: IEEE Trans. Nanotechnol. **3**, 124 (2004)

6

Metallic Nanocrystals
and Their Dynamical Properties

Jens-Boie Suck

Abstract. Metallic nanocrystals have numerous technical applications, especially because of their favourable magnetic and mechanical properties, surpassing often by far those of the corresponding polycrystals. As applies to most of the nano-materials, their importance in applications will grow further and wider in the years to come.

Because of their small extension of only a few nm in all or some of their dimensions, the properties of nanocrystals are generally characterized by two main facts: *a competition of length scales* due to the fact that their extension is smaller than some physical relevant length scale like e.g. the correlation length, screening length, free path length. For small nanocrystals, the network of grain boundaries between the crystallites can occupy up to 50% of the sample volume and thus there is *a strong influence of the network of grain boundaries on the overall properties of nanocrystalline materials.*

After an introduction into the production and characterization methods of metallic nanocrystals and some characteristic properties of their grain boundary network, in this lecture the consequences of both of the two characteristic facts just mentioned for some of the properties of (mainly fcc) metallic nanocrystals are discussed: the melting temperature, the magnetic and the mechanical properties and finally in detail their atomic dynamics .

6.1 Introduction

Nearly all of the preceeding lectures are devoted to non-metallic materials. It seems therefore reasonable to add a chapter on metallic systems, as metals and alloys are another very important class of materials, and metallic nanocrystals play an important role also in technical applications. Nano-materials (**NM**) are characterized by building units, which have one or more of their dimensions (thickness, diameter etc.) below about 100 nm [1, 2]. In this sense nm-particles in an extended matrix (0dim), thin layers with a thickness of less than 100 nm (1dim), needles or lamella with diameters below 100 nm (2dim) and nanocrystals (**NC**) with all 3 dimensions (3dim) smaller than this value belong to the nano-materials.

In general, NM nowadays enjoy an enormous interest and support by rather different communities, as can be seen from their omnipresence in journals and books [3, 4] newspapers [5], TV, and web. The actual and the projected application of NM reaches from strengthening of compound materials like light-weight alloys and polymers [6], packing materials [7], nanostructured devices [8], data storage [9] and new screen materials [10] for computer to technologies and considerable improvements especially of the surfaces of building materials [11, 12], applications in sensor technology, as soft and hard magnets [13], as carriers in medicine [14] and biology [15–17] and unfortunately also weapons [18, 19]. The very promising properties of NM, which lead to an extrapolation of the present world market volume of 10^{11} to 10^{12} Dollars in 2015, resulted in very strong national (German government 2005: 0.3 10^9 Euros) and international support for nano-technologies (**NT**). However, in most recent years one also became aware of the potential problems, which could be connected with the NT, and several committees have investigated the chances but also the (especially also toxic) risks, which could be connected with the production and use of NM [20–25]. Thus the present and certainly also future 'nano-euphoria' will be complemented with some 'nano-scepticism'.

For industrial use the 'Nanotechnology Consumer Products Inventory' has been founded recently and concerning research, the first (10 volume!) 'Encyclopedia of Nanoscience and Nanotechnology' appeared in 2004 [26] and will need and will have its second edition already at the beginning of 2007. Thus NM are not only "in" at present but will be in with increasing importance in the years to come.

Why this excitement and optimism and why this high scientific interest in NC?

The smallness of the particles allows to combine different NC to form new materials, which otherwise cannot or can only be produced with great difficulties and costs, as the nm-materials can be mixed and the grains brought next to each other and react in an un-preceeded manner [27]. Likewise the smallness of the particles allows them to fill spaces and with this change the material properties, especially also to strengthen existing materials.

Even more striking is the fact that at smaller size the diameter of the NC often is below the physical length scales of importance for different physical properties. Examples are correlation and screening lengths or the mean free path for transport properties (some examples will be given below) and therefore nano-materials show *new and sometimes very advantageous properties*, compared to polycrystals, in which the crystallites usually have a diameter of at least 2 to 3 orders of magnitude larger than have the NC. This fact makes nano-materials extremely advantageous for industrial application, where miniaturization of all technical components is desired.

The nm size of the material necessarily leads to an unusual large surface to volume ratio. As a consequence there is a strong contribution from the grain boundary (**GB**) region, i.e. the *crystallite surfaces* and the inter-grain region (**IGR**) to the physical properties of NC. This fact influences strongly their physical properties. Sometimes they can for this reason be viewed as two component systemsconsisting of nanocrystallites and the surrounding grain boundaries with nearly equal volume contribution (see mixing rule below).

On the other hand NC can develop undesirably in strong deformations of polycrystalline materials [28]. In addition, NC are metastable against grain growth, especially when used under extreme conditions [29]. Thus one has to understand the formation conditions and the changes of the material properties if they have to be used under conditions, which favour the, in these cases unwanted, NC or polycrystal formation, respectively, which might lead to material failures. All these different new properties explain the motivation and enormous effort and investment, which are put nowadays into the research, production and improvement of NM.

After this more general introduction into NM, this article will focus on some of the properties and especially on the atomic dynamics of *metallic nanocrystals*. Most of them will have fcc structure and one has to keep in mind that bcc NC might show somewhat different properties in detail [30]. However also NC with aperiodic structure (nanocrystalline quasicrystals) will implicitly be included and mentioned below [31]. For NC of a few nm diameter the volume of the GB may reach up to 30% and more of the sample volume as shown in Fig. 6.1.

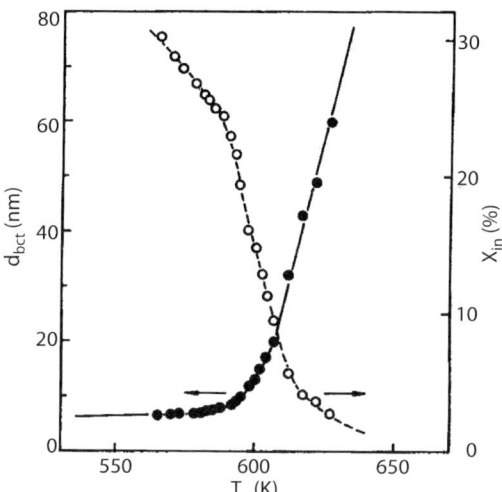

Fig. 6.1. Average size of nanocrystallites and molar fraction of of interfaces as function of the annealing temperature of $Ni_{80}P_{20}$ metallic glass (from [32])

The importance of the contribution of the GB to the physical property K of the sample (density, conductivity etc.) can be estimated very roughly using the *mixing rule* [33]:

$$(K)^n = \lambda_{NC}(K_{NC})^n + \lambda_{GB}(K_{GB})^n \; ; \qquad (6.1)$$

the quantities n (most often near 1), λ_{NC} and λ_{GB} have to be determined separately for each property under consideration, where the λ are the relative volume contributions and can often be estimated from X-ray diffraction pattern. For the property K_{NC} one can use the known values for the corresponding single- or polycrystal in a first approximation. As K is measured, K_{GB} remains as the property to be determined. As examples we mention the use of the mixing rule in estimations of the mechanical properties of NCs [34], of the contribution of the GBs to the Mössbauer spectrum [35] or of the density of the GBs [33]. Based on the facts that the sample density ρ and volume V can be measured, that of the grains is most likely the same ρ_X as that of single crystals of the same material and the volume of the grains, V_X and GB's V_B can be estimated from X-ray diffraction or can be calculated assuming a grain shape and a GB thickness δ, one has

$$\rho = \frac{V_X}{V}\rho_X + \frac{V_B}{V}\rho_B \qquad (6.2)$$

$$\rho_B = \frac{V\rho - V_X\rho_X}{V_B} \; . \qquad (6.3)$$

Because of the large amount of GB-volume, nanocrystalline (**nc**) materials with sufficiently small crystals offer on the other hand the exceptional possibility to study also the properties of GBs. Of this possibility use will be made towards the end of this lecture.

In this article a brief introduction into metallic nc materials will be given followed by some examples concerning some of their properties to demonstrate the influence of crystallite size, d, and GB volume on the properties of NC without any intention for completeness. We therefore recommend to the students to read much more complete review articles like [36–40]. In addition, a large amount of book-titles will be the result of any search in the web. Finally this article will concentrate on the atomic dynamics of metallic nc materials.

6.2 Production of Nanocrystalline Materials

NC are produced in a way, which favours the nucleation of many seed nuclei for crystals, but disfavours crystal growth. The necessity to suppress crystal growth implies that NC are necessarily metastable against crystal growth. Even though crystal growth may be hampered very effectively in many cases, allowing some NC to exist for some time not too far below their melting temperature T_{m} , finally the kinetic barrier to crystal growth will be overcome and polycrystals will be formed with time. If a more general view is taken on nano-materials, as it is done e.g. in the article of Gleiter [36], then also examples of thermodynamic stable NM can be given.

Because of this metastability, NC have to be produced artificially. They can be produced effectively by all methods, which lead to a complete formation of crystals and avoid growth of the initiated crystallization-nuclei beyond the nm size. For applications and also research one guides the production process in a way, which allows to produce nc materials with a well defined *grain size, grain size distribution, chemical composition and purity*. In addition to these parameters, the properties of the GB are decisive for the properties of the nano-material. Often and especially if industrial application is intended, a small amount of elements is added in addition to the elements needed to form the NC and – if embedding of the NC into a matrix is desired – the matrix. This is done to promote a rich nucleation and to hamper crystallite growth. To give an example of an industrially used nc alloy: in the well known FINEMET, $Fe_{73.5}Cu_1Nb_3Si_{13.5}B_9$, the Cu atoms just promote nucleation of the α-FeSi while the Nb in the GB hinders the growth of the crystallites formed [41]. Another recent example is nc MgNi, for which 5 times smaller NC are obtained, if some at% Y are added to the starting material [64].

Below, some of the methods are mentioned briefly, which are commonly used to produce metallic NC.

One of the earliest methods was the production of nc metals or alloys from the **vapour phase**. In vapour phase production the metal, or if nc alloys are the target, the metals are heated up to the vapour phase. The atoms in the vapour are then slowed down by scattering in a He-atmosphere. As a result of this, small clusters or crystals have time to form. To suppress further growth these have to be removed from the growth area by natural diffusion or gas transport onto a cold finger, where they get caught and further growth is suppressed by the low (liquid Nitrogen) temperature (≈ 77 K) and the nanocrystals are collected.

From the cold finger in the middle of the production device the NC are removed. Via a funnel they reach a volume, where they can be compressed under UHV conditions (to remove the rest gas) to form nc material of about a cm diameter and 1 mm thickness [43]. Later gas condensation was also applied to nc ceramic materials, either by direct evaporation or (if the evaporation temperature is too high) by collecting the metallic NC as described before and subsequent oxidation of this material [1]. The cluster size can be influenced by the evaporation and thermalization rate of the atoms. Using these production parameters the grain size (**GS**) distribution can be kept relatively narrow. The resulting grain size distribution often corresponds to a *log-normal-distribution*:

$$\frac{\Delta n}{\Delta(\ln d)} = \frac{1}{\ln \sigma \sqrt{2\pi}} \exp\left\{ - \frac{(\ln d - \ln \overline{d})}{2 \ln \sigma} \right\}, \tag{6.4}$$

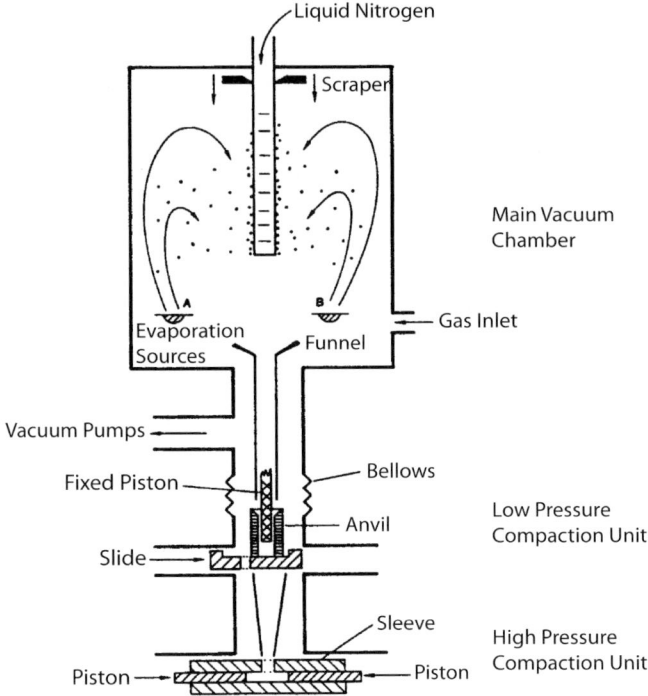

Liquid Nitrogen

Scraper

Main Vacuum
Chamber

Evaporation
Sources

Funnel

Gas Inlet

Vacuum Pumps

Fixed Piston

Bellows

Anvil

Low Pressure
Compaction Unit

Slide

Sleeve

Piston

Piston

High Pressure
Compaction Unit

Fig. 6.2. Gas quenching apparatus for the production of nanocrystals (from H.Gleiter [43])

where n is the number of crystals, d their diameter, $\bar{d}$ corresponds to the mean value of d and σ is the standard deviation. The un-symmetry of this distribution stems from the continuing crystal growth (aggregation of clusters). The compression of the nc material under a pressure of several GPa leads to a densification, which reaches up to 97% of the theoretically attainable density for metallic NC and 75 to 85% for ceramic NC. This implies that the NC must deform under this pressure to reach this high density.

Likewise NC can be produced by **sputtering** [44], again starting from the gaseous phase, which however this time is not produced by heating but by ion bombardment. Using sputtering, the starting material (element or alloy, the latter in a relative concentration, which takes into account also the differences in sputtering effectiveness of different elements) is bonded to the anode and the nanocrystals are formed within the sputtering gas before they are captured at a substrate kept at the required fixed temperature. Both methods have in common that the starting materials have to be very pure including contaminations from gaseous elements like H and that the same condition applies to the gas, in which the NC are formed. These conditions are not always very easily fulfilled and thus, especially at the beginning, gaseous contaminations were not a rare problem of these nano-materials.

Nanocrystals, and this applies especially also to the non-conducting oxide nc materials, can also be produced by **chemical reactions**, where again crystal formation is favoured, but crystal growth is avoided. Pulsed electrodeposition is one of the few methods, with which pore-free metallic NC can be produced [45, 46].

Continuous **electrodeposition** is another method, with which nc samples are produced especially if thin films made from NC are needed [47]. In electrodeposition the plating bath contains the desired metals in solution, while a current of the order of 0.1 A/cm^2 passes between two electrodes, on one of which the nc film will be deposited.

A widespread method to produce also a larger amount of NC is **mechanical attrition** (**MA**). In mechanical attrition NC form mostly in the shear bands of strongly deformed larger crystals. The deformation can be reached by ball milling [48–50] or by direct repeated deformation of the material by bending or rolling [28].

Ball milling is applied to powders of either elements or alloys. Solid spheres of 1–2 cm diameter are accelerated within a thick walled cylindrical vial by rotation of the vial, often in a planetary rotation system (the vials rotate around their cylinder axis and this rotates again around the axis of a disk, on which the vials are fixed). Depending on the rotation speed, which is one of the variable MA production parameters – besides the ratio of ball weight to powder weight, the ball diameters, the duration of the milling, and the temperature –, the balls can reach a very high kinetic energy, with which they impact on the walls of the vial, milling the powder on the walls into finer and finer crystals. The weight of all balls involved has to be 10 to 15 times larger than that of the powder in the vial. To avoid crystal growth at elevated temperatures the self-heating system has to be cooled. The advantage of ball milling is the large amount of sample, which can be produced and the intense mixing of the material, which allows to form e.g. alloys in compositions, which cannot be reached otherwise, e.g. supersaturated alloys. The clear disadvantage is the unavoidable contamination of the sample material by atoms from the vial and ball surfaces and from the gas in the vial. This latter contamination can be reduced by using gas tight vials filled with inert gas. A further disadvantage is the fact that one has less control on the nano-crystallization process and has to accept a broader size distribution. The smallest reachable grain size depends on the hardness of the material, as the working process will start to 'glue' from a certain grain size onwards the NC produced from softer material. The harder the material the smaller are the smallest NC, which can be obtained. Somewhat less contamination seems possible, if instead of a ball-milling rod-milling is used [51].

An efficient method to produce metallic NC, like pure metals, alloy and intermetallics, with very little contaminations is the **hydrogen plasma-metal reaction**. In the hydrogen plasma-metal reaction the metals are arc melted and evaporated in a stream of Ar-H gas (50:50) and the forming nanoparticles are then pushed by the gas flow in the subsequent collecting system [52]. The H in the gas prevents oxidation of the metals at the initial stage of cluster formation and has to be removed afterwards if disturbing.

A more recent method, which rapidly gained widespread use, especially also in industrial applications, is the **nano-crystallization** of metallic glasses, amorphous alloys [53, 54]. Nano-crystallization can be either *nucleation* or *growth* dominated [55]. In the first case, the annealing of the completely amorphous sample leads to the formation of crystallization germs, which grow on further annealing. In the second case, crystallization germs were already produced during the production of the amorphous starting material within the amorphous matrix and these only have to grow when the temperature is increased in the annealing process. To determine the temperature, at which the glass has to be annealed to form a maximum density of nucleation sites, one uses differential scanning calorimetry (see below in Sect. 6.3). The sample is heated rapidly to a constant annealing temperature T_a below the crystallization temperature T_x of the glass. The sample is kept there for a defined amount of annealing time t_a and then heated up to T_x rapidly. T_x (more accurately: the temperature corresponding to the maximum of the enthalpy release, the 'crystallization peak') changes as a function of T_a and t_a, reaching a minimum value, when the maximal possible nucleation density was reached in the amorphous sample [56]. Using this information on the optimal nucleation temperature and time, one succeeds in a complete transformation of the amorphous sample into a nc one. However it is important to realize that the result in general depends on more parameters than just the annealing temperature and time. As the different mobilities of the elements in the starting glass depend on the dependence of the viscosity on temperature, the *fragility* of the glass can be an important parameter [57] as can be the local depletion reaction during crystal growth [58]. A very interesting observation is the correlation between phase separation in the glass on annealing and the NC formation on a *similar length scale* as has the phase separation, possibly one conditioning the other [59].

Most recently some alloys have been found, which can be quenched *directly* quantitatively into the nc phase by **nano-quenching**. Usually this is achieved with a quench rate of about 10^5 K/s, i.e. slightly below the quench rate used to produce corresponding metallic glasses using the melt spinning technique [60]. The advantages of the nano-quenching technique are very similar to those of nano-crystallization, i.e. no larger voids but only the two smaller types of free volumes are produced (see Sect. 6.4), contamination by other elements and additional surfaces due to powder formation can be avoided. However one has to expect that the degree of structural order of the atoms in the GBs is different from that of the nc samples produced by nano-crystallization. Due to the production directly from the melt one has to expect a higher degree of structural disorder in the GBs [61].

6.3 Characterization of Nanocrystalline Materials

Different from those computer simulations (**CS**), in which nc samples are constructed in a systematic manner, e.g. by the Voronoi method [62], experimentally produced nc samples do not grow under permanent control. They therefore have to be analyzed very carefully after their production using *different* methods applied to the same sample. In the following, some of the more frequently used methods to analyze the structure and quality of nc materials are briefly mentioned.

The first method, which is usually applied after production, is **wide angle X-ray or neutron diffraction**. For NC with an upper diameter (in their diameter distribution) of less than 50 nm, wide angle X-ray and neutron diffraction (**WAD**) can give fairly reliable information on the mean size of the NC from the width of the considerably broadened Debye–Scherrer lines as are shown in Fig. 6.3 [63,64].

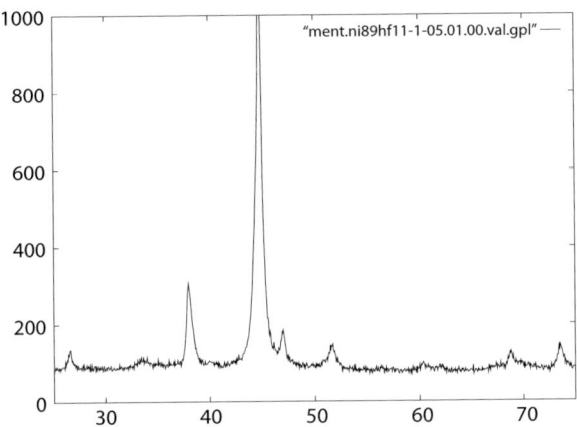

Fig. 6.3. Wide angle X-ray diffraction pattern of nanocrystalline $Ni_{89}Hf_{11}$

This is especially true, if the lines do not overlap and thus each linewidth can be analyzed using – in the most simple approach – the Scherrer formula [65],

$$d = \frac{\lambda}{\cos\left(\theta/2\right)} \sqrt{\Gamma_{nc}^2 - \Gamma_b^2} \qquad (6.5)$$

where λ, θ, Γ are the wavelength of the radiation used, the scattering angle and the widths of the Debye–Scherrer peaks of the nc and polycrystalline (bulk (b)) sample, respectively. More reliable results are obtained in using the Warren–Averbach method, which allows to separate the contribution from the smallness of the diffraction volume, i.e. the NC, and stresses [66]. In addition to this more general information, the WAD-pattern, in the representation of the static structure factor, S(Q), can be used to calculate the radial distribution function, g(r), from which the number of nearest neighbours and their mutual distances can be obtained [67].

Nanocrystalline material is best and most reliably characterized by high resolution **transmission electron microscopy (HRTEM)** [68–70] as shown in Fig. 6.4 for nc NiHf. Taking patterns from different regions of the nc material, one can also determine the distribution function of the diameters by transmission electron microscopy.

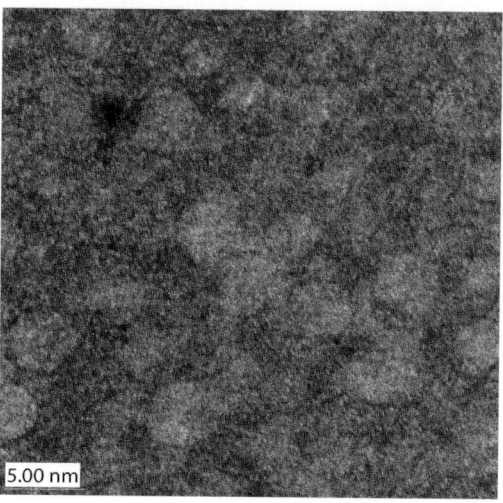

Fig. 6.4. TEM pattern of nanocrystalline $Ni_{89}Hf_{11}$ produced by melt spinning [71]

One should however keep in mind that the diameters obtained via HRTEM and WAD will most likely be different, as the linewidth of the diffraction peak is given by the size of the volume for coherent scattering, while TEM gives a 2dim projection of the particles [72].

The short range order, SRO, of the atoms can be studied using Extended X-ray Absorption Fine Structure, **EXAFS**, as has been done already rather early in the investigation of nc materials [73]. In this method the fine structure of the X-ray absorption, measured as a function of the X-ray energy well behind the absorption edge, is separated out. After Fourier transformation, one obtains the radial distribution function, g(r), for small to medium large r. As the method is element sensitive, because of the characteristic backscattering-amplitudes of each element, one additionally obtains information on the partial occupancies of sites by the elements next to the ad-atome, from which the electronic wave started, which is backscattered by the surrounding atoms.

Another local probe, which has been used repeatedly [74,75] is **Mössbauer spectroscopy**. Mössbauer spectroscopy can be applied, if the sample contains

Mössbauer nuclei like ^{57}Fe or ^{119}Sn. As absorption is a *local* event, information on the distribution of these nuclei, e.g. preferred within nano-crystallites or dominantly in the IGR, can be obtained.

For NC at surfaces, scanning probe microscopy like (**STM**) and (**AFM**) [76–78] can be used very efficiently to determine the diameter and diameter distribution of the NC at the surface.

Due to the fact that at smallest wavevector transfer, Q, scatterers with largest diameter, d, will predominantly contribute to the measured diffraction intensity,

$$d = \frac{2\,\pi}{Q} \tag{6.6}$$

scattering contributions from the NC as a whole can be investigated effectively by X-ray and Neutron Small Angle Scattering (**SAXS** [79, 80] and **SANS** [81, 82]). This method can be used to determine the *radius of gyration* (mean diameter) of the NC using Guinnier plots (ln(SAS-intensity) as a function of Q^2). However in this analysis of the small angle diffraction pattern one has to make an important assumption on the dominant shape of the scattering units, the NC.

Due to the importance of the GB in nc material and with this the distribution of free volume, **positron-lifetime** measurements have been applied to nc material [83]. In positron-lifetime measurements the sample is irradiated with positrons and the time until annihilation with an electron in the sample is measured. This gives information on the electron density and with this on the free volume. In a simplified manner: the shorter the lifetime, the higher the local electron density, the smaller the free volume without electrons.

A *most* important method in the analysis of the quality of nc materials is the **thermal analysis** of the sample. This is done most often using Differential Scanning Calorimetry, **DSC**. About 5 to 20 mg of the sample, enclosed in a container, are heated in a furnace, where an empty container, identical to the one containing the sample, is fixed in the same furnace. The difference in heating needed to keep both containers (with sample and empty) at the same temperature is measured as a function of temperature (dynamic scan) or of time (isothermal scan), revealing the thermal properties of the sample, e.g. like the specific heat (if correctly calibrated). This analysis is of primary importance for NM, especially if these are produced by nano-crystallization or nano-quenching, as it will prove or disprove the presence of a secondary phase, like remaining amorphous parts of the sample, an unexpected further crystalline phase, segregation of elements from the NC into the GB region [84] or show phase transitions with increasing or decreasing temperature [85].

6.4 Some General Properties of Grains
and Grain Boundaries of Nanocrystals

Metallic nc materials are made up of NC separated by the IGR, which may be either just the GB region with a width of normally less than a nm [86] or the desired matrix, into which the NCs are embedded. The width δ of this region may become of primary importance for the properties of the GB network, especially if the mean grain size d approaches δ [87]. The structure and physical properties of the GB network, which fills the space between the NCs, depends strongly on the method used to produce the nc material. If experimentally slow growth like in nano-crystallization or in CS systematic construction is at the origin of the material, the GB network does not seem to be very different from that of polycrystals [88]. If the nc material is produced by a rapid quench from e.g. the liquid metal or alloy like e.g. in nano-quenching and some CS, one has to expect a considerable amount of disorder in the GB network (see model in Fig. 6.5). This will influence the physical properties of this part of the sample volume and these properties will *change* after structural relaxation, as is shown below at the example of NiHf in Sect. 6.6.

The dominant part of the GB does not seem to be made up by small angle GB [61] and for monoatomic NC the thickness of the GB is of the order of 0.5 to 1 nm, the width of the region with a statistical distribution of atoms being at least 0.4 nm [86]. The distribution of atomic distances in this region is rather broad, making diffusion of atoms *within this region* easy.

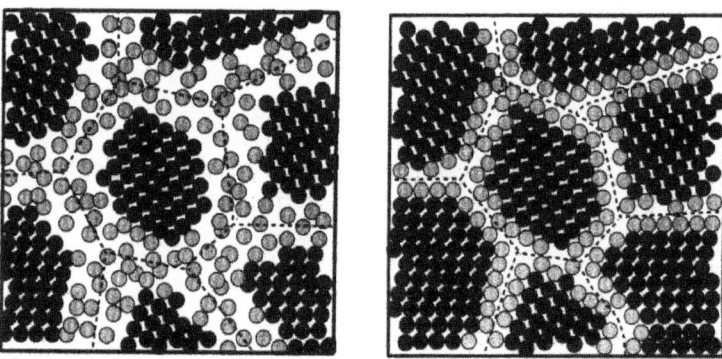

Fig. 6.5. Model of NC and GB of a monatomic nc material

Experimentally it has been found that the dislocation density can be very high in non-equilibrium GBs [89] and their concentration may be very different within the GBs and the grains [90]. Other experiments suggested a kinetically induced, locally disordered GB-network [91]. Independent of the fact that GBs may be nearer or farther away from thermal equilibrium, some disorder in the GB network cannot be avoided, as also in slow growth NC normally grow without correlation in their mutual orientation, which then will lead to disordered regions of some atomic diameters between them in the IGR [43]. The grains are largely unaffected by the production method and their properties resemble approximately those of polycrystalline grains, if they are not influenced by external stresses. Interaction between neighbouring grains is very common.

Concerning the relation between d and δ, there seems to be a *smallest size* for NCs, below which nc materials do not form any longer. Instead, amorphization of the samples is observed, when d approaches approximately a value of $2\,\delta$ [92–94]. Interestingly, there also seems to be an upper size limit, above which solid state amorphization cannot be achieved any longer [95]. Besides the instability of nc material against decreasing grain size there are other stability limits. One of them is the instability against grain growth, which may be very different for different nc materials. For many of them grain growth is very weak at temperatures below $0.3\,T_m$ [96], where T_m is the melting temperature of the nc material (see Sect. 6.5.1), but increases rapidly at higher temperatures. The main source of this instability is the high excess free energy in the GB, which can be reduced by grain growth, because then the ratio of the GB-volume to grain-volume is reduced, as has been shown by grain size dependent thermodynamic experiments [97, 98].

An important role in connection with this excess free energy is played by the *free volume* especially in the GB region. Positron lifetime experiments have found three different types of free volume in nc material: 1. mono-vacancy size, 2. micro-voids of 10–15 vacancies and 3. large voids [83]. Of these three types, only the first two are found in nc material produced e.g. by electrodeposition, nano-crystallization and nano-quenching. The amount of free volume provides another intrinsic instability limit, as has been shown by thermodynamic calculations [99]. On a similar basis a limiting existence temperature has been predicted for nc materials [100].

In spite of these sources of instability, to which the high surface energy of the NCs contributes as well, the stability of NC against growth may under certain circumstances be impressively high. If the size distribution is small, Ostwald ripening, the growth of the larger grains at the cost of the smaller ones to reduce the surface energy, is disfavoured. In addition, neighbouring grain surfaces are often parallel, thus do not provide a clear gradient with a growth direction for the grains [33]. Both obstacles can be overcome by grain rotation, after which coalescence of small grains to form a larger one is possible [101, 102]. Finally grain growth is determined by diffusion, the velocity of which is different for different elements. It sets in as soon as for one of the elements involved the root mean square displacement reaches the diameter of the crystallites. For NC made up of alloys with several components, the sometimes large diffusion distances necessary to form the stoichiometric alloy can be a further obstacle for crystal growth [58].

Nanocrystalline materials, formed first as a loose assembly of clusters (grains) kept together by van der Waals forces, can be compressed to a maximal reachable density, which is considerably larger than the 78% of the bulk density, the limiting density when packing of hard spheres is assumed. Thus the grains are obviously deformed under compression without leading to texture formation. At the same time the porosity of the nc material decreases, especially if the compression is performed at temperatures above room temperature (RT).

As a final remark concerning general properties and the structure of NCs and GBs the possibility that the GB network may have a fractal dimension of 2.4 instead of 3 [103] should be mentioned, as this would influence the spectral dimension of the density-of-states of the atomic vibrations in the IGR region.

6.5 Some Examples of the Special Properties of Nanocrystals

In this section a few examples are given for properties, which are influenced by smallness of the NC or the much enhanced surface to volume ratio in comparison with larger crystals. In nearly all cases, a *competition of length scales* is responsible for the special properties of NC in comparison with the 'traditional' properties of polycrystals.

6.5.1 Melting Temperature

Both, the melting temperature, T_m, and the cohesive energy characterize the strength of the bonding in solids: the larger the cohesive energy the higher T_m. Heterogeneous melting starts from defects on the surface: the larger the surface to volume ratio the lower is T_m. One can start the description of the melting temperature of nano-materials from different starting points, however in all cases the small diameter of the grains and, caused by this, the much enhanced surface to volume ratio and the surface properties will enter the calculation of the decreasing melting temperature T_m with decreasing diameter of the NCs.

One possibility is to start from the condition that the chemical potential, μ, of the solid and the liquid should be equal at T_m. Developing μ up to second order in temperature, T, and pressure, p, and expressing the corresponding derivatives of μ by thermodynamic parameters, using the Gibbs–Duhem relation $-Vdp + SdT + md\mu = 0$, one arrives at a change of T_m relative to the normal bulk (b) melting temperature, T_b, $\Delta T_m = T_b - T_{NC}$. This depends on the grain radius, r_s, the densities, ρ, the surface tensions, γ, and the molar latent heat, L_m [104, 105]

$$T_b - T_{NC} = \frac{2}{L_m \, \rho_s \, r_s} \left[\gamma_s - \gamma_l (\frac{\rho_s}{\rho_l})^{2/3} \right] T_b \tag{6.7}$$

In (6.7) the suffix s characterizes the solid NC and l the liquid.

If one starts from the cohesive energy of the nanocrystal either via the liquid drop model [106] known from nuclear physics or directly [107, 108], the decreasing T_m results from the fact that with decreasing grain size the *relative* number of less bound (in the model often half bound) atoms (with radius r_a) on the surface of the grains increases rapidly, leading to a decreasing T_m according to

$$T_b - T_{NC} = \frac{2\,r_a}{r_s}\,T_b \tag{6.8}$$

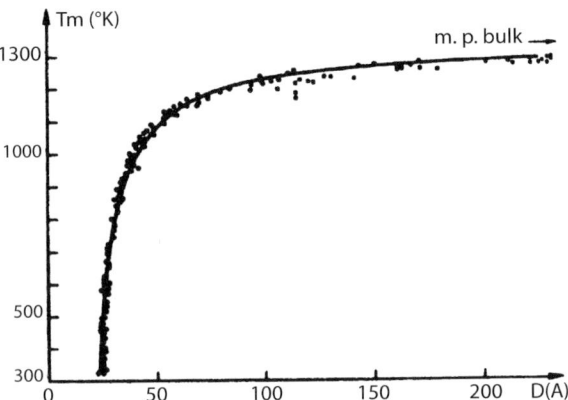

Fig. 6.6. Melting temperature of nm-size gold particles in dependence of the particle diameter. The line is the description of the experimental results by (6.7) [104]

Figure 6.6 shows the drastic decrease of T_m with decreasing gold particle size, especially for NC with $d < 5$ nm, and the very reasonable description of the experimental results by (6.7).

6.5.2 Some Magnetic Properties of Nanocrystalline Materials

While for melting the large surface to volume ratio is decisive, for the magnetic properties it is the smallness of the NC diameter, d, compared to the extension of the magnetic exchange interaction, which causes very favourable magnetic properties, having found numerous technical applications [109]. The often used statement: *tailoring material properties via nano-crystallization* applies especially to the magnetic properties. In fact, soft magnetic materials, required for small losses of alternating currents in e.g. coils and transformers, can be produced in the same way as hard magnetic materials for e.g. magnetic storage devises [110]. To give just two examples: nc $Fe_{79}Zr_7B_9$ is a soft magnet with $\mu_0 \times H_c = 10^{-7}$ Tesla, where μ_0 is the magnetic field constant and H_c the coercivity. If one replaces the 7 at% Zr by Nd one has a very hard nc magnet with $\mu_0 \times H_c = 10$ Tesla [111]. Many of these materials and also the just mentioned ones are obtained by nano-crystallization.

This production technique gives the additional advantage that by choosing the appropriate annealing temperature and time one can produce NC in an amorphous matrix, which will have different (normally softer) magnetic properties and may in some technically important cases be chosen in a way that the magnetostriction, λ_{m}, of the NC and of the surrounding matrix compensate each other: $\lambda_{\mathrm{av}} = 0$ [112–114], which is another example of tailoring the properties in a very favourable way. The industrially on large scale – like the large coils of the new proton accelerator in J-PARC (Japan) – applied nc material, FINEMET [113–115], $Fe_{73.5}Cu_1Nb_3Si_{13.5}B_9$, mentioned already in Sect. 6.2, is an example for this: at 17 at% Si the α-FeSi crystallites with $\lambda = -6\,10^{-6}$ just compensate the large positive $\lambda = 20\,10^{-6}$ of the amorphous matrix. From the mixing rule given above (see (6.1)) it follows that the volumetric weight of FeSi has to be 70 to 80% for this compensation to occur. Another big advantage of this production method is the fact that more complicated geometries can be formed as long as the starting material is a metallic glass, being therefore very ductile and easy to deform [113]. After nano-crystallization the material keeps the previous shape without breaking in spite of the fact that it becomes brittle as a consequence of the thermal treatment. This kind of nc material is stress-insensitive, which polycrystals even with $\lambda_{\mathrm{av}} = 0$ are not [113]. A further advantage is the fact that nano-crystallization can be performed within a magnetic field, which again can tailor the hysteresis in a favourable way [113]. For the final properties not only the size of the grains but also their shape plays an decisive role.

In a ferromagnetic polycrystal each crystallite has an easy crystallographic direction for the magnetization to align, the *magnetic anisotropy* characterized by K_1, the bulk *anisotropy constant*. As the lattices of different crystallites are normally oriented at random with respect to each other, the magnetic orientations are also at random outside a strong magnetic field. In principle the same situation applies also to nc materials as long as the grains are large (e.g. 100 nm) and/or the inter-grain region is non-magnetic and thus hinders the magnetic interaction between neighbouring grains. In this case one deals with isolated nc magnetic regions. However the situation changes drastically, if the nc are small and the inter-grain material is transmitting the magnetic coupling. The important length scale here is the the *exchange correlation length*, L, compared to the grain size, d. Differently speaking, the critical scale is, where the exchange energy, characterized by the *exchange stiffness constant*, A, starts to balance the anisotropy energy

$$L \sim \sqrt{\frac{A}{K_1}}. \tag{6.9}$$

The exchange correlation length depends on the material, as (6.9) shows. For Co it is only about 6 nm, for Fe 20 to 40 nm. If $d > L$ the local easy magnetic direction will win and the magnetism in each nanograin is determined by its own easy direction as in polycrystals. If $d < L$ the ferromagnetic exchange interaction will win and the magnetization will not follow the easy direction of the individual grains, i.e. *the local anisotropy is averaged out* and instead of K_1 one has to use an *effective* coupling constant [113]

$$\langle K \rangle \approx K_1 \left(\frac{d}{L} \right)^6 \sim \frac{K_1^4}{A^3} \, d^6 \, . \tag{6.10}$$

For this to happen, the region (matrix) between the grains, often only with a thickness of 2–3a [116], where a is the mean atomic distance in this region, must not block the magnetic interaction, showing the importance of the inter-grain region. Thus for $d < L$ the effective anisotropy energy is reduced by coupling N grains together and favours soft magnetic properties. N is usually calculated as for cubic geometry

$$N = \left(\frac{L}{d} \right)^3 , \tag{6.11}$$

which leads to a rapidly varying number of coupled grains: if $L = 35$ nm, a very common value of L for Fe containing materials, $N = 5$ for $d = 20$ nm, and $N = 42$ for $d = 10$ nm. Obviously there is an optimum grain size for *soft* magnetic properties, which is around or below 10 nm [117, 118]. For *hard* magnetic materials, as they are needed for data storage on nanoscale, these facts are most important as well: one has to make sure that there is *no* coupling across the inter-grain region to guarantee that the information stored in one grain is not altered by the information stored in the neighbouring grains [111].

As is shown in Fig. 6.7 at the example of the coercivity, the very favourable low values of H_c and $1/\mu$, the inverse of the permeability, at small grain size increase very rapidly $\sim d^6$ until they reach a maximum between 100 and 200 nm from where they decrease linearly with increasing d as the would do in polycrystals as well.

Most of these properties can be described reasonably well using the *random anisotropy model*, RAM, originally developed to describe the magnetism of metallic glasses (with $d = a$) [119], or – more adaptable to the specific situation of nc grains in a GB-network or in a matrix with different magnetic properties – with its extension, the *generalized random anisotropy model*, GRAM, [120, 121]. In these models, the average anisotropy $\langle K \rangle$ results from the mean fluctuation amplitude of the anisotropy energy within the sub-volume $V = L^3$, which is self-consistently related to $\langle K \rangle$

$$L \sim \sqrt{\frac{A}{\langle K \rangle}} \tag{6.12}$$

$$\langle K \rangle = \frac{K_1}{\sqrt{N}} \, . \tag{6.13}$$

With help of the saturation magnetization, J_s, with $J_s = v_g \, J_s^g + v_i \, J_s^i$ and g and i meaning the volume weighted contributions from the grains and the

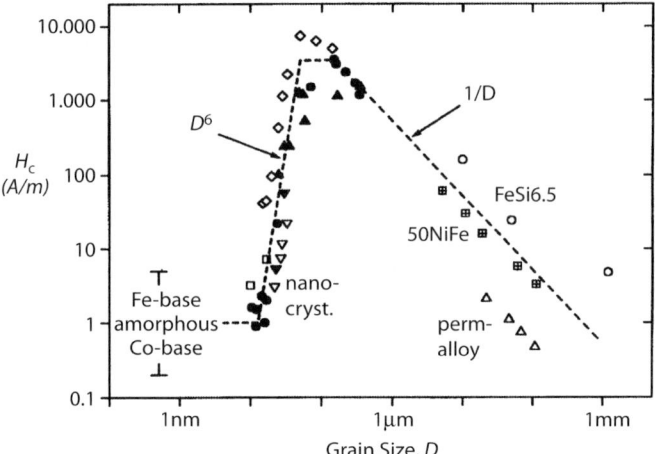

Fig. 6.7. Grain size dependence of the coercivity [117]

IGR or matrix, respectively (see (6.1)) and two dimensionless weighting factors, p_c and p_μ, with values between 0 and 1, one can then calculate the most important properties for applications, the coercivity, H_c, and the permeability

$$H_c = \frac{p_c}{J_s} \frac{K_1^4}{A^3} d^6 \tag{6.14}$$

$$\mu^{-1} = \frac{\mu_0}{p_\mu J_s^2} \frac{K_1^4}{A^3} d^6 \ . \tag{6.15}$$

In the GRAM one has to take into account, that the exchange stiffness A of the grains of diameter d, may be different for the inter-grain region, A_i, of width δ. The *effective exchange stiffness*, A_{eff} is then [116, 120, 121]

$$A_{\text{eff}} = \frac{A(1 + \delta/d)}{(1 + (A\,\delta)/(A_i\,d))} \ . \tag{6.16}$$

These examples demonstrate that going from polycrystals to NC provides a large flexibility to adjust the magnetic properties of the material for a wide region of applications.

6.5.3 Some Mechanical Properties of Nanocrystalline Materials

As was mentioned above correspondingly for magnetic NC, also in the case of mechanical properties one can say that often nc materials are made to improve these properties, i.e. to *improve the strength and hardness of metallic polycrystals* [52]. In polycrystals deformation leads to a shearing of crystal planes within each crystallite. This produces dislocations and proceeds via a continuous nucleation of dislocations from Frank–Read sources *inside* the grains. Dislocations are line defects in the regular crystal lattice and Frank–Read sources are based on the decay of larger dislocations into smaller ones, because the energetic cost of two smaller dislocation is smaller than that of a large one [122]. Frank–Read sources of diameter d_s need stress to operate. The stress required increases proportional to $1/d_s$ [123, 124]. As the source has to be inside the grain, i.e. $d_s < d$, the dislocation production via Frank–Read sources becomes more and more difficult with decreasing grain size, as the stress needed to bow out a dislocation finally reaches the theoretical shear stress limit [125]. Dislocations can move through the grains on well defined slip systems and can interact with each other [123, 126]. GB hinder the transmission of dislocations, which then leads to a *dislocation pile up* as also observed in CS [127] and to a corresponding hardening of the material. This makes nc materials so attractive also from their mechanical point of view, because this effect is used to improve the hardness and yield strength of materials by producing them as nc material instead of polycrystals.

The smaller the grains are and the more important the network of GB becomes, the more pronounced is this dislocation pile up and with this the hardening, known since a long time as Hall–Petch (H–P) mechanism [128, 129]. The dependence of the nucleation stress, σ_n, on grain size, d, is according to Hall and Petch

$$\sigma_n = \upsilon_0 + \frac{K}{\sqrt{d}}. \tag{6.17}$$

Here σ_0 and K are the single crystal yield stress and a constant related to the strength of the GB [130] , respectively. This $1/\sqrt{d}$-dependence has been followed down to grain diameters of 10 to 40 nm in experiments [131] and CS [123]. The actual value of the smallest d, at which the hardness still increases, the *'strongest size'* as it is called [132, 133], depends on the stacking fault energy, γ, of the material under consideration [130]. This is the misfit energy caused by atomic planes stacked out of sequence [134]. As has been shown most recently in a plasticity map [130, 134] the larger γ the smaller is the diameter, down to which the Hall–Petch relation holds. To give an example, the CS [130] show that in Al, which has a large γ, at $d = 24$ nm the normal dislocations mechanism just described still perfectly works, while at the same grain size in Pd (small γ) complete dislocations can no longer be nucleate inside the grain. More accurately, the stacking fault energy alone is not sufficient for predicting correctly the mechanical behaviour of the nc metal, but the *generalized stacking fault energy function* (a double maximum curve) has to be known, involving the stable (in the minimum between the maxima) and the unstable stacking fault energy (on one of the maxima) [135].

It is again a very impressive prove for the abundant situations of *length scale competitions* in the physics of NC that there exists a *smallest size d_m* down to which the H–P relation holds and below which a *reverse Hall–Petch mechanism* is observed: at room temperature and below with decreasing d the

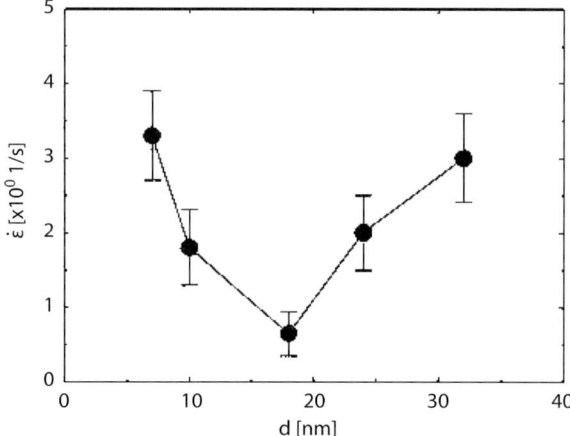

Fig. 6.8. Strain-rate as a function of the grain size. The minimum is observed where the material is hardest (strongest size) [136])

material becomes *softer* again as is shown in Fig. 6.8 at the example of the strain rate, which is an indication for the resistance of a material against deformation and with this for its hardness.

The reason for this change in the mechanical properties of nc materials is a change in the way the materials reacts on the deformation. This change becomes necessary because of the smallness of the grains, which does not allow any longer the production of complete extended dislocations *inside* the grains. What happens at grain sizes below d_{m} depends, as discussed above, on the stacking fault energy (curve): For high γ there is a cross-over transition from perfect dislocation slip to a GB mediated deformation, i.e. when the *intra*grain dislocation slip is no longer possible, the deformation activity is transfered to the GBs and an *inter*grain deformation takes over, which has been called GB-diffusion-creep, a coupled GB-sliding [137] and GB-diffusion [130], which allows the grains to slip against each other. In detail, the GB-diffusion-creep involves a significant amount of discrete atomic activity like a *stress assisted free volume migration* by a sequence of atomic hoppings and uncorrelated atomic shuffling, which seem to be the rate controlling processes for GB-sliding [61, 138]. This cross-over causes the increasing weakening of the resistance of the nc material against deformation, called the *reverse Hall–Petch* mechanism.

For nc materials with low γ, between the two regions just discussed, a transition region exists, in which the grains are too small for a complete *intra*grain activity, i.e. for Frank–Read sources to operate in the usual manner and the dislocations are therefore created in the GB. However at this size the grains are still too large for the final *inter*grain mechanism. In this region extended partial dislocations are formed, preferably at the GBs, which transect the entire grain leaving behind stacking faults, which hinder dislocation propagation.

Thus a softening of the nc material as observed in the reverse H–P mechanism does not take place yet. At higher temperatures than discussed up to now, i.e. at $0.7\,T_\mathrm{m}$, GB sliding due to GB diffusion (Coble creep [139]) is observed.

For the transition from the H–P region to the reverse H–P region, which has been observed first in CS [132, 133] also several experimental results exist. One of the most recent results is shown in Fig. 6.9 for nc Ni and Ni–W alloys [47].

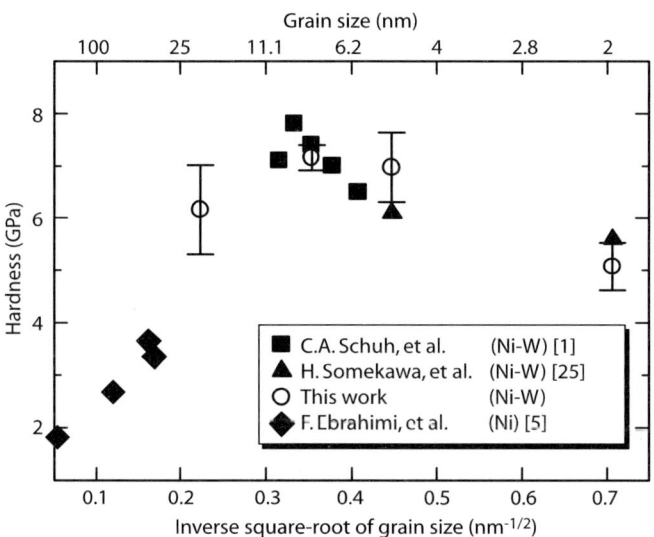

Fig. 6.9. Hardness of electrodeposit nanocrystalline Ni–W alloys as a function of grain size: the maximum indicates the 'strongest size' [47])

Interestingly also the 'memory' concerning deformations works differently in NC and in polycrystals. While deformations leave in the latter a *residual dislocation network*, which causes a broadening of the diffraction peaks also *after* tensile deformation, this is not the case in nc material: experimentally and by CS it has been shown for nc Ni that the diffraction peak broadening is *reversible* [125, 140].

Even though quite an amount of information from experiments on the mechanical properties of nc material exist [131, 141–145] and also from some model calculations [146], most of these detailed results have been obtained in CS, which due to the limited simulation time have to use very strong and fast deformations. Thus their results have to be regarded as being more on one side of the deformation scale.

6.6 Vibrational Properties
of Metallic Nanocrystalline Materials

After having given examples from three different properties of nc materials, their melting temperature, their magnetic and mechanical properties, which are all affected and changed either by the very high amount of IGR and grain surfaces or by the small diameter of the crystallites, we now turn to the atomic dynamics of metallic NC.

6.6.1 General Remarks on the Atomic Dynamics
   of Metallic Nanocrystals

Because of their metallic character, inelastic neutron and X-ray scattering (**INS** or **IXS**) are the most often used technics for the study of the atomic dynamics of metallic NC. However, when discussing the results from the (in most cases) INS experiments on metallic NC done in the past 10 to 15 years and the corresponding computer simulations on NC with radial atomic interactions, one should not overlook that the first (to our knowledge) systematic investigations of the influence of crystal size (down to 5 nm) and of relaxation on the atomic dynamics (as far as reflected in Raman scattering spectra) were done 25 years ago on Si-particles [147] and have included CdSe [148], CeO_2 [149], SnO_2 [150], InP [151], TiO_2 [152] in recent years. The shift of the Raman line with grain size is very systematic and has even been used to determine a crystallite diameter [149].

   In inelastic scattering experiments one determines the intensity of the scattered radiation as a function of the wave vectors k_0 of the incident and k of the scattered particle (neutron or photon in this case). From these quantities one determines the *energy transfer*, $\hbar\omega$, and the *wave-vector transfer* Q or the momentum transfer, $\hbar Q$, respectively. As nc materials and polycrystals are isotropic samples, one only needs to determine Q instead of Q in this context.

$$\hbar\omega = \frac{\hbar^2}{2M}\left(k_0^2 - k^2\right) \tag{6.18}$$

$$\hbar Q = \hbar(k_0 - k) \tag{6.19}$$

$$Q = \sqrt{k_0^2 + k^2 - 2\,k_0\,k\,\cos\theta} \tag{6.20}$$

with M the mass of the scattering unit and θ the scattering angle, under which the intensity was recorded.

   After correcting and normalizing the scattered intensity (see. e.g. [153] for the case of neutrons) from INS or IXS (including nuclear resonance scattering at synchrotron radiation sources) experiments, one determines either the *vibrational density-of-states* (**VDOS**), $F(\omega)$, or the *dynamic structure factor*, $S(Q,\omega)$, in all cases, where there is no contribution from magnetic scattering and all scatterers in the sample have identical scattering properties. If this latter condition is not fulfilled, i.e. different scatterers (e.g. elements) contribute with different weights to the measured intensity, then one measures *generalized* functions, a weighted sum over all *partial* contributions from each of the different scattering units. The weights are given by the 'strength' of the coupling of each scatterer to the probe. In the case of the dynamic structure factor it is the *total dynamic structure factor*, which is the weighted sum over the *partial dynamic structure factors*, $S_{i,j}(Q, \omega)$,

$$\sigma_{sc} \, S(Q, \omega) = 4\pi \sum \sum b_i b_j \sqrt{c_i c_j} S_{i,j}(Q, \omega)$$
$$+ \sum \sigma_i^{inc} c_i S_i^s(Q, \omega) \quad (i, j = 1...n). \tag{6.21}$$

σ_{sc} and σ^{inc} are the total and the incoherent scattering cross-sections and S_i^s is the self part of the dynamic structure factor of the element i. In (6.21) it was taken into account that the sample might also scatter incoherently and thus $S_i^s(Q, \omega)$ was added to to total dynamic structure factor.

Instead of the VDOS one determines the *generalized vibrational density-of-states* (**GVDOS**), $G(\omega)$, from the measured spectra, if several different scatterers contribute to the measured intensity. The GVDOS is again the weighted sum of the *partial density-of-states*, $g_i(\omega)$, of each element in the sample, weighted by the strength of the coupling of this element to the scattered probe (here for neutrons again)

$$G(\omega) = \frac{\sum w_i c_i g_i(\omega)}{\sum w_i c_i} \tag{6.22}$$

$$w_i = \frac{e^{-2W_i} \sigma_{sc}^i}{M_i}. \tag{6.23}$$

Here e^{-2W_i} is the Debye–Waller factor and M_i is the mass of the scatterer i. The density-of-states, as always, contains the information on the *single particle* dynamics of the system, while the dynamic structure factor reflects also the *collective* atomic dynamics for coherent neutron scatterers, as is needed for e.g. measurements of dispersion relations. It provides information on the *wavelength-dependence* of the atomic dynamics.

For nc materials with GS below approximately 20 nm a complication comes in due to the fact that the samples are still isotropic but no longer homogeneous. Thus the measured intensity of the scattered probes stems from three different kind of sample parts: the grains, their surfaces and the IGR-network. We know from the previous sections that for NC with diameters below approximately 10 nm all of them can contribute with similar weight (scattering volume). In addition, while the grains themselves and approximately also their surfaces may be near thermodynamic equilibrium, the GBs may contain strains and disorder, which make their properties, as already discussed in Sect. 6.4, very dependent on the way the nc material was produced and on its thermal history. While in CS it is easy to separate out contributions from these three different regions [154], it is very difficult to separate them out reliably in experiments. Thus one aim of investigations of the atomic dynamics of nc materials with different methods is the separation of at least two of these different contributions if possible.

As always, the knowledge of the VDOS (or taking in its place the GVDOS in lack of the VDOS) allows then to calculate in the harmonic approximation different thermodynamic properties, like the *vibrational part* of the Helmholtz free energy, F, of the internal energy, E, of the entropy, S, of the specific heat, c, and the Debye cut-off frequency or Debye temperature, Θ_D, using the dimensionless energy transfer β with k_B the Boltzmann and R the gas constant, which together with T determine the dimensions of the quantity [155–157]:

$$\beta = \frac{\hbar\omega}{k_B\,T} \tag{6.24}$$

$$g(\beta) = k_B\,T\,g(\hbar\omega) \tag{6.25}$$

where $g(\hbar\omega)$ stands for either $F(\hbar\omega)$ or $G(\hbar\omega)$.

$$F_{vib} = 3RT \int_0^{\beta_{max}} \ln(2\sinh(\beta/2))\,g(\beta)\,\mathrm{d}\beta, \tag{6.26}$$

$$E_{vib} = 3RT \int_0^{\beta_{max}} \frac{(\beta/2)^2}{(\sinh(\beta/2))^2}\,g(\beta)\,\mathrm{d}\beta \tag{6.27}$$

$$S_{vib} = 3R \int_0^{\beta_{max}} [\beta/2\,\coth(\beta/2) - \ln(2\sinh(\beta/2))]\,g(\beta)\,\mathrm{d}\beta \tag{6.28}$$

$$c_v^{vib} = 3R \int_0^{\beta_{max}} \frac{(\beta/2)^2}{(\sinh(\beta/2))^2}\,g(\beta)\,\mathrm{d}\beta \tag{6.29}$$

$$\Theta_D = T \sqrt[n]{\frac{n+3}{3}\,\langle\mu^n\rangle} \tag{6.30}$$

$$\langle\mu^n\rangle = \int_0^{\beta_{max}} \beta^n\,g(\beta)\,\mathrm{d}\beta \tag{6.31}$$

Likewise on can calculate with help of the VDOS (GVDOS) mean values of dynamical quantities like the mean force constant, f, for a monoatomic sample, and the Debye–Waller coefficient, $2W$, or the mean value of the vibrational amplitude, u:

$$\overline{f} = \frac{\hbar^2}{(k_B\,T)^2} \sum_n f_n = M \int_0^{\beta_{max}} \beta^2\,g(\beta)\,\mathrm{d}\beta \tag{6.32}$$

$$2W(Q) = \frac{\hbar Q^2}{2\,M\,k_B\,T} \int_0^{\beta_{max}} \frac{\coth(\beta/2)}{\beta}\,g(\beta)\,\mathrm{d}\beta \tag{6.33}$$

For crystals with cubic structure one can separate out the mean square vibrational amplitude, u,

$$2W(Q) = \frac{1}{3} Q^2 \langle u^2 \rangle \tag{6.34}$$

$$\langle u^2 \rangle = \frac{3 \hbar^2}{2 M k_B T} \int_0^{\beta_{max}} \frac{\coth(\beta/2)}{\beta} g(\beta) \, d\beta \tag{6.35}$$

6.6.2 Phonon Confinement

One of the first observations made in studying the atomic dynamics of NC was the *smoothness* of the spectra measured in INS experiments [51]. This has been observed in *all* subsequent cases (see Fig. 6.10 as a drastic example). This reduction of structure in the spectra (mainly the van Hove singularities) has been interpreted as being due to the reduction of the *mean free path* of the phonons within the small grains, i.e. *phonon confinement* [72, 158–162]. Calculating the number of cycles, noc, for a phonon of frequency ν with sound velocity v starting at one side of a grain with diameter d

$$noc = \frac{\nu \, d}{v} \tag{6.36}$$

one finds for a phonon with a frequency of 6 THz and a sound velocity of 5000 m/s just 12 cycles until it reaches the other side of its 10 nm cage. This limitation in path length corresponds to a shortening of the lifetime, τ, of the phonon

$$\tau = \frac{\hbar}{\Gamma} \tag{6.37}$$

where Γ corresponds to that part of the linewidth of the excitation, which is caused by the finite life time due to the finite path length. As the GVDOS of the NC can be well described by a convolution of the GVDOS of the poly-crystal with a damped harmonic oscillator function, D_ω, at each frequency ω [159–161]

$$D_\omega(\omega') = \frac{1}{\pi \, \omega} \frac{\Gamma}{(\omega/\omega' - \omega'/\omega)^2 + \Gamma^2} \tag{6.38}$$

and no dependence of the GVDOS on temperature was found between 30 K and RT [160], making contributions from un-harmonic vibrations to the smoothing of the structures not very likely, it seems reasonable to assign the dominant part of the observed broadening to the life time limitation due the small diameter of the crystallites.

A corresponding loss of structure in the spectra has also been observed in CS [93, 164–166], however has not been discussed in detail. Repeatedly it was mentioned that especially the longitudinal phonons are concerned by this damping. However one has to keep in mind that in all cases either metals or binary alloys have been investigated, where the van Hove singularity for the

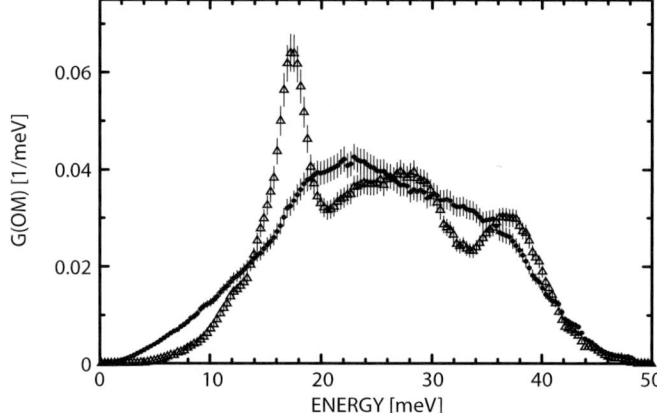

Fig. 6.10. The generalized vibrational density-of-states of nanocrystalline (small-nano as.quenched) and polycrystalline $Ni_{89}Hf_{11}$. The strong loss of structure of the GVDOS is obvious [163]

longitudinal acoustic modes (and of most optic bands) is the most pronounced and sharpest feature in the VDOS, which makes damping effects more obvious and spectacular than for the normally broader distribution of transverse acoustic modes. An example for the grain size dependence of the damping is given below in Fig. 6.14.

6.6.3 Grain-Size Dependence of the Atomic Dynamics

As the contribution from the IGR and the surfaces of the NC to the scattering volume varies with the diameter of the nanocrystals, the change of the atomic dynamics with changing GS is the obvious first source of information on the varying contributions from IGR, surfaces and grain interior. One normally measures the spectra in comparison with that of a polycrystalline sample of (if possible) the same chemical composition and atomic structure and one assumes that for polycrystals the contribution from the GB region, which should be two to three orders of magnitude smaller than in the case of NC, can be safely neglected. In CS one often compares with the dynamics of a single crystal, which in essentially all cases is found to be rather similar to that of the grain interior, except for the length of the free path and a possible compression of the metallic nc grains [93, 166]. This contraction can be quantified in measuring the lattice constant of the grains. In two cases, the additional ('excess')intensity found above the upper frequency limit ω_{max}^{pc} of the VDOS of the corresponding polycrystal (pc) has been attributed to the vibration of the atoms in the grain interior under compression. A more likely interpretation of this intensity is however that it is due to the broadening of the VDOS due to phonon localization in the grains. Experimentally it is difficult to distinguish these two possible origins of the high frequency modes, as decreasing grain size will lead to stronger localization *and* could lead to higher compression (which then would have to be quantified), increasing the high energy intensity in both cases.

The variation of the GS can be done in different ways. Normally the next larger GS is obtained by annealing the sample with the smaller GS. This however represents a severe limitation on the interpretability of the measured spectra, as the difference observed (in all corresponding experiments so far) is due to the grain growth *and* the structural relaxation in the GBs during annealing. For samples prepared by nano-crystallization this effect is reduced, as all relaxations, the activation energy of which belongs to the lower part of the distribution of activation energies, already take place when the sample with the smallest GS is made. The most reliable way of changing the GS without influencing too strongly the structural state of the GBs, is to produce different samples with different GS from the very beginning, though even this procedure requires a change of the production parameters, which might influence the structure of the GBs as well. As the technique for getting information on the contribution from the GB to the spectra is obvious, most of the experiments and CS have investigated also or mainly the GS-dependence of the atomic dynamics [72, 93, 154, 158, 160–162, 166–168].

As can be seen from Fig. 6.10, apart from the smoother structure and some intensity at energies above $\omega_{\mathrm{max}}^{\mathrm{pc}}$, the main characteristic of the VDOS of nc materials is the low energy intensity in the VDOS and in the dynamic structure factor. This had already been seen in the first INS experiments on *nanocrystalline quasicrystals* like icosahedral PdSiU [169] and AlCuV [170]. However the nanocrystalline nature of the samples has only been proven by HRTEM after the interpretation of the experimental results [171]. Even though it has been suggested that these low energy modes (**LEM**) could be due to inter-grain vibrations [72], in nearly all of the investigations by experiments and CS this low energy intensity and the corresponding measured 'excess' specific heat [172] is attributed to the interface between grains, i.e. the grain surfaces (surface modes) [166, 173] and the atomic vibrations in the IGR-network [99, 100]. A considerable contribution from surface modes on crystals of nm size (and below) to the low energy intensity in the VDOS of the particles and correspondingly to their low temperature specific heat had already been found in a theoretical investigation very early [174–176] similar to what had been predicted by calculations for clusters of much smaller size [177]. One therefore expects that a variation of the GS should lead to a change of the *intensity* of the low energy part of the VDOS (and of the dynamic structure factor), which for spherical crystallites should be proportional to $1/d$. Unfortunately, experimentally both quantities are rather inaccessible at lowest energies, because this energy region is occupied by the huge foot of the peak of the elastically scattered particles. A *reliable, quantitative* separation between this foot of the elastic line and the inelastic spectra is in most cases only possible at about four times the FWHM (full width at half maximum) of the elastic peak [178]. One normally bridges the thus emerging gap towards $\omega = 0$ by an appropriate function,

$$g(\omega) = x \, \omega^e \tag{6.39}$$

For 3dim polycrystalline samples e is normally 2, because the VDOS is determined at lower energy by the linear dispersion branches of the acoustic modes, which leads, following Debye, to a dependence proportional to $\omega^{\dim-1}$ of the lowest part of the VDOS. Here dim is the dimension of the sample and in all cases investigated so far dim $= 3$. If the surfaces of the grains can be regarded as one atom thick layers, dim $= 2$ and the VDOS of the surface modes should start as a linear function.

Turning now to the results from experiments and CS, all investigations of the GS dependence of the LEM in NC come quite generally to the conclusion that the low energy intensity of the VDOS *increases with decreasing d*, i.e. increasing surface to volume ratio and/or increasing GB to grain volume. This already is a very strong argument for *assigning the origin of the LEM to the surfaces and the IGR-network*. With one exception, all results also show a *decrease of this LEM intensity and the corresponding specific heat on annealing*, as this leads to a release of internal strains [158] and some loss of free volume and an increase of the structural order in the GBs as sketched in the model in Fig. 6.5.

Less unanimous is the discussion of the dependence of the initial slope of the VDOS on ω and of the scaling function. Likewise diverse the question is discussed, which quantity actually scales with d. As far as surface modes are concerned, all results from CS and from experiments agree on a linear increase of $F(\omega)$ at smallest energies (see Sect. 6.6.5 below). If all three partial contributions are present: atomic vibrations in the grain interior (GI), surface modes (SM) and IGR modes, each with its proper weight

$$G(\omega) = A \, g_{\mathrm{GI}}(\omega) + B \, g_{\mathrm{SM}}(\omega) + C \, g_{\mathrm{IGR}}(\omega) \,, \tag{6.40}$$

then in most cases a dominant ω^2-dependence of the GVDOS has been found experimentally [72, 159–162] and in CS [93, 165] even if its initial slope had been modeled with both contributions

$$g(\omega) = \alpha \, \omega + \beta \, \omega^2 \tag{6.41}$$

The scaling with GS again is not unanimously reported. Several experiments investigating the GS-dependence of the VDOS (GVDOS) conclude that β in (6.41) scales with d^{-1}, as does the density of GB or the number of atoms in the IGR [72, 159, 160, 162]. However also indications for a scaling of α/β with d^{-1} has been reported [161].

6.6.4 Comparison with the Atomic Dynamics
of Metallic Nanocrystals
and the Related Amorphous Solids

Metallic NC can not only be obtained from metallic glasses (by nano-crystalli-zation, see Sect. 6.2), but become instable at a critical grain size (see Sect. 6.4 above), which for amorphization is near 1 or 2 nm. This has been investi-gated in CS [93] and has been observed also experimentally [94]. The atomic dynamics of both systems have been studied and compared with each other in CS [93, 164] and in experiments on nc and glassy $Ni_{80}P_{20}$ [168, 179] and $Ni_{33}Zr_{67}$ [167]. In all cases, the GVDOS of the metallic glass was definitely more intense at lowest energies than that of the corresponding NC (see Fig. 6.11).

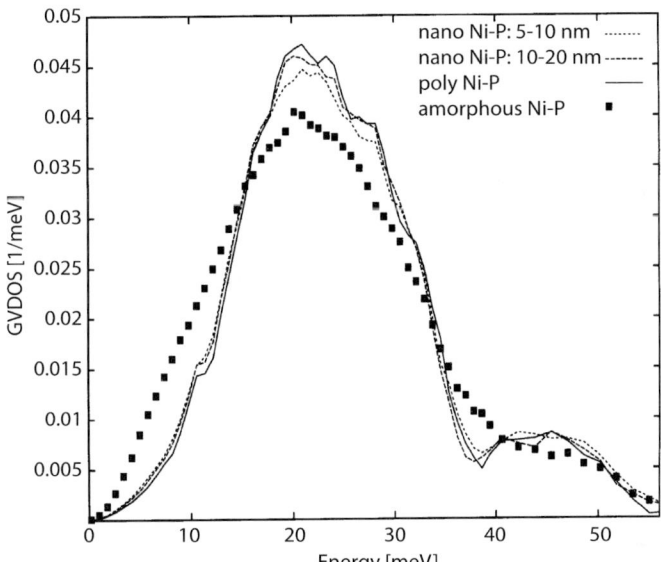

Fig. 6.11. Generalized vibrational density-of-states of the metallic glass $Ni_{80}P_{20}$ and of the nanocrystalline sample with grain sizes between 5 and 10 nm made by nano-crystallization from the glass. The higher intensity of the GVDOS of the glass at energies below 15 meV is clearly seen [163]

Correspondingly, also the vibrational part of the specific heat, calculated using the GVDOS in place of the VDOS in (6.29), of the glass is always found to be higher than that of the nc sample at low temperatures, as is shown in Fig. 6.12.

It has to be mentioned that these (repeated) results for the vibrational part of the specific heat are different from the results obtained in calorimetric

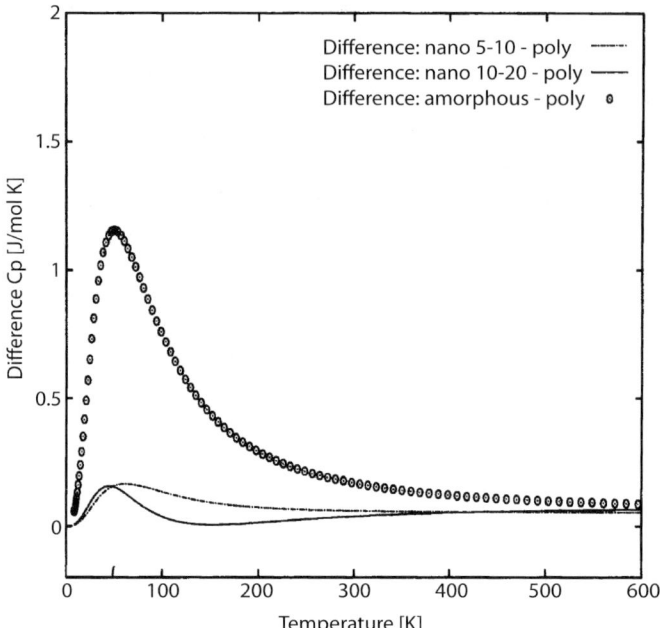

Fig. 6.12. Difference of the vibrational part of the specific heat of the metallic glass $Ni_{80}P_{20}$ and of the nanocrystalline sample with grain sizes of 5 to 10 nm and of 10 to 20 nm made by nano-crystallization from the glass. The corresponding specific heat of the polycrystal is subtracted in both cases. The higher 'excess' specific heat of the glass is clearly seen [163]

measurements of the specific heat of NiP, where higher values were measured for the nc sample than for the starting metallic glass [180].

6.6.5 Specific Investigations Concerning the Contribution of the Grain Boundaries and Surfaces to the Observed Spectra

Besides these more general investigations, some experiments and CS have been done to answer specific questions. The *difference* of the the VDOS of nc Ni, produced by inert gas condensation with and without compaction, and the weighted VDOS of coarse-grained Ni has been determined ($G_{NC} - G_{PC}$) [181]. This difference VDOS is localized in an energy region between the lowest accessible energy (about 1 meV) and 17 meV with a maximum near 10 meV. The exponent e (see (6.39)) of the starting slope of the difference VDOS is only slightly larger 1. Within the severe limitations of this experiment due to a H/H_2O contamination of 2.5 at% no further details of the difference function could be analyzed.

An intended H-content of 4 at% and below (to avoid H precipitation) in nc Pd, made by inert gas condensation and compaction, was used to study the atomic dynamics of the interface region, i.e. outer and inner surfaces of the grains and the IGR [182]. On the basis of a previous study by INS [183], it was assumed that the H atoms are dominantly bound at outer and inner surfaces, which certainly applied to the Pd black sample. As Pd has a very small incoherent scattering cross-section, while H scatters predominantly incoherently, the scattering from Pd and H could be separated by separating the incoherent and coherent neutron scattering via polarization analysis. While the description of the scattering from Pd required a VDOS starting proportional to ω^2, as one would expect for a compact 3dim crystal, the local DOS (LDOS) calculated from the incoherent scattering from H started nearly linear for both nc samples. This is in accordance with the assumption that the surfaces (especially in Pd black sample) are the origin of these vibrations, and it also suggests that in this case also the vibrations of the atoms in the IGRs contribute with an intensity of similar energy dependence.

These experimental results are very well supported by CS [166, 173–176] on a single crystallite of several nm diameter, where bulk and surface modes could be investigated separately and lead to different slopes of the LDOS. In the investigation of two nc samples, Cu and Ni, by CS [154] all three contributions (see (6.40)) have been separated, however only the slopes of the VDOS due to the vibrations of the atoms within the grains and due to the atoms in the GB region (surfaces + IGR) have been analyzed, giving exponents of 2 for the crystallites and 1.5 for the interfaces. The fact that the surface atoms contribute the intensity at lowest energies and the steep slope of this partial VDOS leaves however the possibility that the surface-VDOS starts proportional to ω, as one would expect from the dimensionality. This most detailed CS also provides additional informations concerning the atomic dynamics in the three regions under discussion: the partial VDOS of the atoms *within* the grains is very similar to the VDOS of the polycrystal and is *not GS dependent*. The VDOS of the atoms in the GB (at $d = 5$ nm about 36% of all atoms are in the GBs) is a very much smoothed and broadened version of the VDOS of the atoms in the gain interior, very similar to the VDOS of an amorphous alloy. Due to this broadening its intensity is higher at low and at high energies than that of the bulk VDOS and it extends to higher energies as it also does in the case of metallic glasses [184]. However the *shape* of this VDOS changes little with GS, which suggest that the *atomic structure* of the GBs in this model, constructed using the Voronoi method, does not change very much with GS, while their *amount* (which is normalized-out in the VDOS) increases proportional to $1/d$ (for spherical grains).

Most recently the *dynamic structure factor*, $S(Q, \omega)$, was determined for two nc samples in a continuous region of energy and momentum transfers for the first time. This allows to investigate not only the energy but also the *wavelength dependence* of the scattering and provides thus additional information. The two nc $Ni_{89}Hf_{11}$ samples with different grain size were made by

nano-quenching (see Sect. 6.2). This method avoids inner surfaces like in powders (e.g. from MA) and larger voids, as they are found in most samples
produced by other methods [83]. One sample material consisted of spherical
NC with diameters between 5 to 8 nm (small nano), the second were lamella
structures and needles with a diameter of 5 to 10 nm and a length of 100 to
400 nm and some spheres of 3 to 10 nm (large nano) [71,185]. Part of each of
the two samples materials was used to induce a gentile structural relaxation
in the GBs by a short annealing (1 h at 773 K and 823 K, respectively) without detectable grain growth. However a grain growth of less than 10% would
not have been detected in the HRTEM characterization of the four samples.
For comparison, a polycrystalline sample was produced from the small-nano
material. An annealing of 4 days at 1073 K was necessary to grow a polycrystalline sample with grain diameters of 250 to 500 nm. This latter sample
was used as reference. As an example for the dynamic structure factor of
a nc-sample in comparison with that of the corresponding polycrystal, both
$S(Q,\omega)$ are shown in Fig. 6.13.

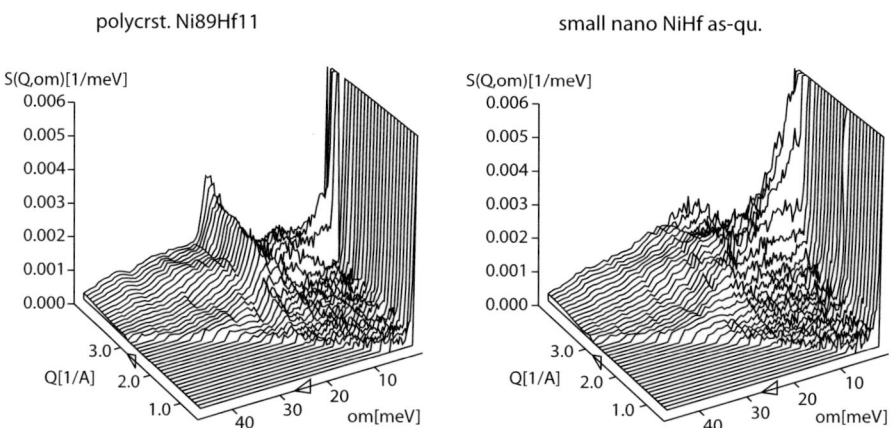

Fig. 6.13. The total dynamic structure factor of polycrystalline (*left*) and nanocrystalline (small-nano as-quenched) (*right*) $Ni_{89}Hf_{11}$ measured at RT with an incident
energy of 7 meV (FOCUS, SINQ). The smoothing of the structure in the spectra
(due to the limitation of the lifetime of the vibrations in the NC) and the strong
amount of LEM next to the Debye–Scherrer peak are obvious

The comparison of the two figures shows two important results: The spectra of the NC (small-nano) at $Q = $ const have lost their structure present
in the same kind of spectra of the polycrystal. As has been discussed above,
this broadening of the structures is interpreted as being due to the limited
free path of the excitations in the small NC, vibrational confinement. This
interpretation is corroborated by the observation that the sample with the
larger NC (large-nano), produced by nano-quenching as well, shows a slightly

small nano NiHf ann.

large nano NiHf ann.

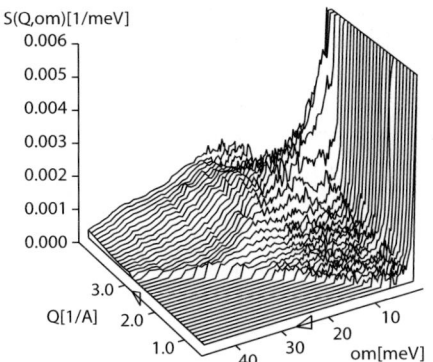

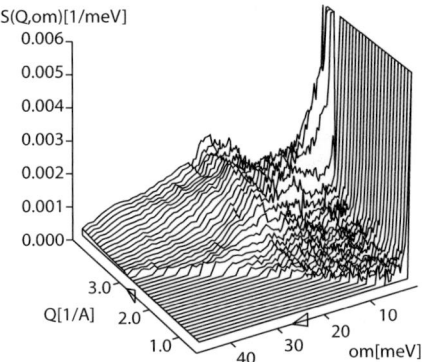

Fig. 6.14. The total dynamic structure factor of nanocrystalline $Ni_{89}Hf_{11}$ after annealing for 1 h at 823 K (small-nano *left*, large-nano *right*) measured at RT with an incident energy of 7 meV (FOCUS, SINQ). The more pronounced structure of the spectra (at $Q = $ const) and reduction of the amount of LEM next to the Debye–Scherrer peak in the sample with the larger grain size (right) are obvious

less strong damping, as is demonstrated in Fig. 6.14, most likely because the phonon pathway is extended in one of the directions.

The intensity of the LEM, discussed above also, does **not** increase uniformly in Q-space proportional to Q^2, as localized vibrations would do in the harmonic approximation, but is very strong next to the Debye–Scherrer peaks covered in this INS-experiment. This shows that the low energy intensity is to a large extend due to propagating modes with small wavevector q, observed here in Umklapp scattering. The strong concentration next to a reciprocal lattice 'point' (ring because of polycrystal) suggests that a considerable part of these modes are of transverse type.

The fact that the annealing was done without detectable grain growth, gives the possibility to separate out the change introduced by structural relaxation in the atomic dynamics of the nc material using the difference $\Delta S(Q, \omega)$ of the measured dynamic structure factors before and after annealing. One expects that structural relaxation will happen mainly in the 'liquid like' GBs (after the rapid quench) [61] and considerably less on the grain surfaces. In addition, there is no reason to assume that the extremely broad (and therefore in the figures invisible) quasi-elastic magnetic scattering from Ni should change on annealing, and this therefore should drop out quantitatively in the *difference* dynamic structure factor. Therefore the annealed-out intensity will be dominated by modes stemming from the IGR component (interface-modes), while the intensity *remaining* after structural relaxation will contain vibrations from the now more ordered IGR and the *surfaces* of the grains (surface-modes). The energy dependence of these two components (as-quenched minus annealed and annealed minus polycrystal) is compared in

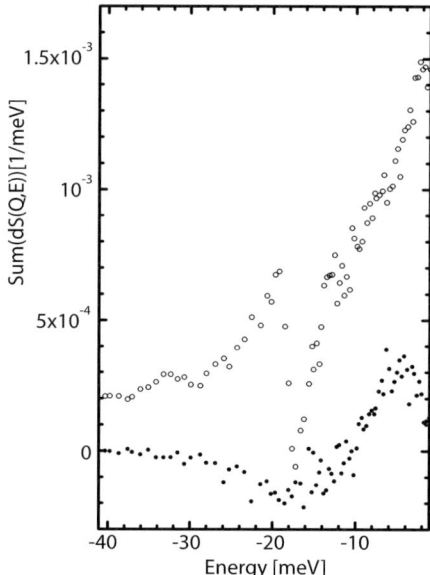

Fig. 6.15. Sum (25 to 36 1/nm of cuts at Q = const through the difference of the total dynamic structure factors of nanocrystalline $Ni_{89}Hf_{11}$ (small-nano: as-quenched – annealed (*filled circle*), annealed – polycrystal (*open circle*) shifted up by 2 10^{-4} to separate the spectra). The very different character of the spectrum annealed out (GB-modes) and remaining after structural relaxation (GB, mainly surface modes) is very clearly demonstrated [186]

Fig. 6.15, where the sum of adjacent cuts (at Q = const) through $S(Q, \omega)$ in the region of the first Debye–Scherrer peak is show. There the LEM intensity is strongest.

This comparison clearly demonstrates that the annealed-out modes, the dominant part of which most likely stems from the IGR region, (interface-modes) have a *different* energy dependence compared to the remaining modes, where the *surface modes* most likely will dominate. The interface-modes have a positive difference between 1.1 (smallest reachable energy) and 11 meV. This intensity disappears on annealing and is transfered to some extend to higher energies, as the negative difference in Fig. 6.15 shows. The most important result however is that the annealed-out modes have their intensity maximum at 5 meV while the surface mode intensity continues to increase towards $\omega = 0$ within the energy resolution of the experiment. The observed difference is in good agreement with the results of the most recent CS, where also the surface modes dominate the intensity at lowest energies [154], as is the case here. A very similar intensity transfer has also been observed in the structural relaxation of metallic glasses [184].

Another similarity with metallic glasses, though less obvious, comes from the comparison of the Q-dependence of the LEM, which are annealed out.

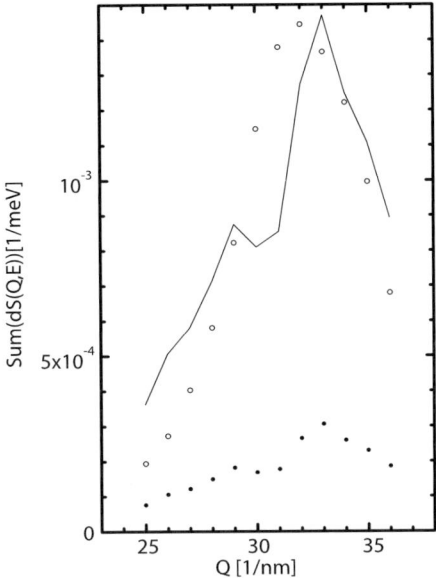

Fig. 6.16. Sum (1.13 to 11.11 meV) of cuts at $\hbar\omega = $ const through the difference of the total dynamic structure factors of nanocrystalline $Ni_{89}Hf_{11}$ (small-nano:
as-quenched – annealed (*filled circle*), annealed – polycrystal (*open circle*)). The
broader Q-dependence of the spectrum annealed out (GB-modes) than that of the
remaining modes after structural relaxation (interface, mainly surface modes) is
demonstrated by the line, the annealed-out modes normalized to the remaining
modes [186]

The dependence of these two $\Delta\ S(Q,\omega)$ on wavevector transfer (Q-dependence) is different, as is demonstrated in Fig. 6.16, where the LEM *inelastic structure factor* (integrated from 1.13 to 11.11 meV) is shown: the surface modes are sharply concentrated around the Debye–Scherrer peak, as
one would expect for phonons, the longitudinal and transverse dispersions
of which start at the reciprocal lattice point (Γ point). The interface-modes
have a broader distribution around the peak, as is demonstrated by the two
different intensities normalized to the same peak hight.

The same kind of spectra, as have just been discussed for the dynamic
structure factor can also be investigated for the *difference GVDOS*,
$G_{asqu} - G_{ann}$ and $G_{ann} - G_{poly}$, as is shown in Fig. 6.17.

The intensities at energies below 3.1 meV of these $G(\omega)$ have been replaced by a Debye spectrum. The very different character of the GVDOS of
the interface-modes and of the surface-modes is evident. The good agreement
between these results and those of a recent CS corroborate strongly our as-
signment of the spectra to interface- and to surface-modes, which could be
investigated independently of each other in the CS [154]. At the same time
the differences shown in Fig. 6.15 and 6.17 demonstrate (on absolute scale)

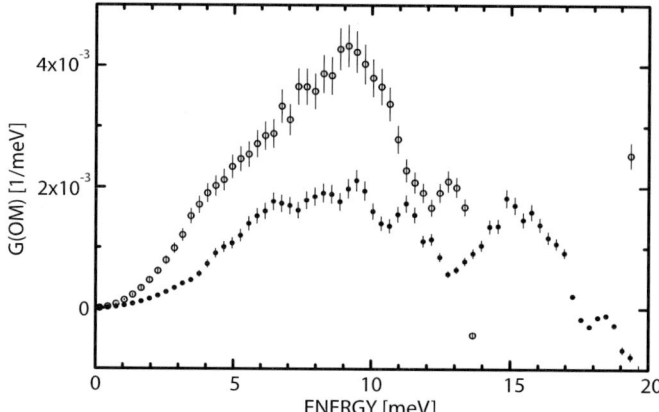

Fig. 6.17. Differences of the generalized vibrational density-of- states of nanocrystalline (small-nano) $Ni_{89}Hf_{11}$: $G_{asqu} - G_{ann}$ (*filled circles*) and $G_{ann} - G_{poly}$ (*open circles*). The intensities of $G(\omega)$ below 3.1 meV have been replaced by a Debye spectrum. The strong difference between the GVDOS of the annealed-out and the remaining modes is most obvious [186]

that the spectrum 'lost' on annealing represents a *considerable* part of the remaining low energy spectrum even without detectable grain growth.

This now gives a hint on the rather different results concerning the *slope* of the measured GVDOS. If not only a special partial VDOS has been determined as just discussed, the VDOS (GVDOS) will always be determined by a superposition of three contributions stemming from the interior of the grains (IG), the surface of the grains and internal surfaces e.g. from voids and holes (SG) and the IGR network (IGR), each with its proper weight (see (6.40)). For NC with diameters below 10 nm the surfaces and the IGR contribute with a considerable volume to the scattered intensity, increasing with decreasing grain size.

As mentioned above, the contribution from the interior of the grains will be proportional to ω^2 at lowest energies and its energy dependence can be regarded as being unchanged in zeroth approximation, as no GS-dependence of this contribution has been observed in careful CS [154]. As the intensity of this contribution varies proportional to the total crystallite volume, i.e. with GS, the spectra (or VDOS) of the coarse-grained polycrystalline sample have to be properly weighted to achieve a quantitative subtraction of this contribution. The surfaces should contribute an intensity proportional to ω at lowest energies, as in this case the spectral dimension $(\dim - 1) = 1$. The intensity of this contribution will strongly depend on the GS, i.e. the *surface* to volume ratio, and will be important only for small grains or large surface/volume ratios. The intensity of this contribution, varying opposite to the variation of the contribution from the interior of the grains, can therefore only be regarded as essentially unchanged, if samples with the same GS and porosity are compared. There will be some surface relaxation on annealing, however if filling of internal holes can be excluded during annealing, this effect will be smaller than the changes observed in the structure and internal strains of the GB-network. It is therefore very likely the last contribution in (6.40), which will vary most strongly with production conditions and thermal history. The intensity of this contribution will be proportional to ω^e, where $1.4 \le e \le 2$, as a reduced dimensionality of the IGR contribution cannot be fully excluded [103]. A large compressibility of the GB-network measured via Mössbauer spectroscopy and its decrease after annealing [75] and a reduction of strain on annealing by factors up to *four* have been reported from a Warren–Averbach analysis of the WAD pattern [158, 160]. Rather ordered and very disordered, – characterized as 'liquid like' [61, 164] –, GB structure have been discusses, which enables large changes as a consequence of annealing. Our own results demonstrate a strong loss of LEM on annealing without detectable grain growth, which we attribute mainly to the structural relaxation in the IGR [186]. The changes observed in the spectra with decreasing (or increasing) GS is therefore to a large extend caused by the *changing of the weighting functions* A, B, C in (6.40).

It depends therefore strongly on the sample and its thermal history, which of the three (or two, if the VDOS of the polycrystalline sample was subtracted) contribution in (6.40) will dominate the slope of the total VDOS (GVDOS) at lowest accessible energies:

$$g(\omega) = a\,\omega^2 + b\,\omega + c\,\omega^e \qquad (6.42)$$

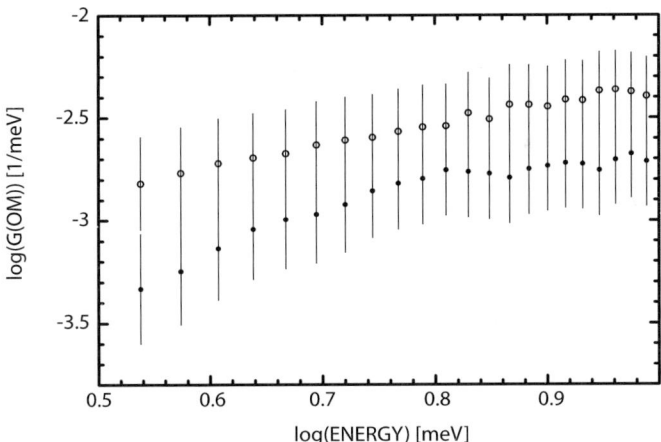

Fig. 6.18. Log-log representation of the low energy part of the differences of the generalized vibrational density-of- states of nanocrystalline (small-nano) $Ni_{89}Hf_{11}$: $G_{asqu} - G_{ann}$ (*filled circles*) and $G_{ann} - G_{poly}$ (*open circles*) above 3.1 meV. The difference in the initial slopes by a factor of 2 is clearly seen [186]

If d is large, the first term in (6.42) stemming from the vibrations within the grains will be dominant. If d is small, say below 10–15 nm, the second and the third term will start to dominate the *low* energy VDOS. Only in this case one should observe a clear deviation from the ω^2-law.

As the surfaces should contribute proportional to ω and the interior of the grains proportional to ω^2 with similar weight for *small* NC, it should mainly depend on the contribution from the IGR, which overall dependence will be observed. If the IGR-network is strongly disordered, it may contribute with $e < 2$, similar to metallic glasses [184]. The total $g(\omega)$ will then show a similar dependence. If the sample is well relaxed, e will be near 2. If then the contribution from the surfaces is not the dominant one, one will observe an ω^2-dependence within present experimental accuracy.

If one of the interfacial partial contributions can be separated out, as has been done for the surface modes in CS [166, 173–176] and experiment [182], and here for that part of the GB modes, which can be annealed out, the specific ω-dependence of this partial contribution can be determined.

This interpretation agrees in fact very well with the results presented above: if one investigates the slopes of the two difference GVDOS, which are shown in Fig. 6.17, above 3.1 meV in a double log presentation, one finds an ω^2 dependence for the GVDOS of the modes assigned to the IGR (annealed-out modes) between 3.1 and 6 meV (see Fig. 6.18). For the modes assigned mainly to the surfaces the GVDOS shows a linear dependence on energy between 3.1 and 9.3 meV. For the VDOS of all GB modes (not only the annealed-out part of them) Derlet and collaborators found an initial slope of $e = 1.5$ [154].

Summing up one has to conclude that presently there is no simple answer to the question: which slope will the VDOS of NC have at low energies: the resulting slope depends on the relative contributions of the three partial DOS shown in (6.40). The relative weight of each of them depends on the relative scattering volume from the grain interior, the grain surfaces and the interface region, i.e. on grain size.

References

1. R.W. Siegel: Synthesis and Processing of Nanostructured Materials. In: *Mechanical Properties and Deformation Behaviour of Materials Having Ultra-Fine Microstructures*, NATO ASI Series, Series E: Appl.Sci. vol 233), ed. by M. Nastasi, D.M. Parkin, H. Gleiter (Kluver, Dordrecht 1993) pp 509–538
2. R.W. Siegel: Physics Today **46** 64 (1993)
3. Materials Today **9** 1-64 (2006)
4. http://www.aspbs.com
5. T. Ahmia: die Tageszeitung **7970** 17 (2006)
6. J. Jordan, K.I. Jacob, R. Tannenbaum, M.A. Sharaf, I. Jasiuk: Mater. Sci. Engineer. A **393** 1 (2005)
7. http://www.k-zeitung.de April (2006)
8. http://www.heise.de: c't **9** 20 (2006)

9. Mater.Magn.Data Storage **31** vol.5
10. H. Chander: Mater. Sci. Engineer. R **49** 113 (2005)
11. VDI-Nachrichten 3.2.06
12. http://www.baulinks.de/webplugin/2006/
13. M.E. McHenry, D.E. Laughlin: Acta Mater. **48** 223 (2000)
14. http://nano.cancer.gov/news-center/
15. Angw.Chem.Intern.Ed. **45** 3165 (2006)
16. D. Geho et al.: Bioconjugate Chem. **17** 654 (2006)
17. J.K. Herr, J.E. Smith, C.D. Medley, D. Shangguan, W. Tan: Anal. Chem. **78** 2918 (2006)
18. J. Altmann: Military Nanotechnology: Potential applications and preventive arms controll (Abingdon New York 2006)
19. J. Altmann, M.A. Gubrud: Military, Arms Control, and Security Aspects of Nanotechnology. In: *Discovering the Nanoscale*, ed. D. Baird (Amsterdam 2004)pp 269–277
20. http://idw-online.de/pages/de/news159009
21. CRN Global Task Force on Implication and Technology April (2006)
22. *Nanotechnology, Assessment and Perspectives* ed. by G. Schmidt et al. (Springer, Berlin 2006)
23. http://allpr.de/43180
24. http://www.pressetext.ch/pte.mc?pte=060506002
25. http://www.handelsblatt.de/pshb/fn/relhbi/sfn/buildhbi/cn/
26. *Encyclopedia of Nanoscience and Nanotechnology*, ed. by H.S. Nalwa (Academic Press, London, San Diego 2004)
27. H. Mori, H. Yasuda: Mater. Sci. Forum **269–272** 327 (1998)
28. J. Schiotz: Mater. Sci. Engin. A **375–377** 975 (2004)
29. G.J. Fan, L.F. Fu, D.C. Qiao, H. Choo, P.K. Liaw, N.D. Browning: Scripta Mater. **54** 2137 (2006)
30. Q. Wei, S. Cheng, K.T. Ramesh, E. Ma: Mater. Sci. Engineer. A **381** 71 (2004)
31. A. Inoue: Adv. Engin. Mater. **3** 669 (2001)
32. K. Lu, R. Lück, B. Predel: Acta Metall. Mater **42** 2303 (1994)
33. C.E. Krill, R. Birringer: Mater. Sci. Forum **225–227** 263 (1996)
34. D.A. Konstantinidis, E.C. Aifantis: Nanostr. Mater.**10** 1111 (1998)
35. U. Herr, J. Jing, R. Birringer, U. Gonser, H. Gleiter: Appl. Phys. Lett. **50** 472 (1987)
36. H. Gleiter: Acta mater. **48**, 1 (2000)
37. A. Inoue, A. Takeuchi: Mater. Sci. Engin. A **375–377** 16 (2004)
38. A.L. Greer: Thermodynamics of Nanostructured Materials. In: *Mechanical Properties and Deformation Behaviour of Materials Having Ultra-Fine Microstructures*, NATO ASI Series, Series E, Appl.Sci. vol 233, ed. by M. Nastasi, D.M. Parkin, H. Gleiter (Kluver, Dordrecht 1993) pp 53-77
39. A.L. Greer: Mater. Sci. Forum **279-272** 3 (1998)
40. A.L. Greer: Physical Phenomena in Fine Layered Structures. Chap.12 in: *Comprehensive Composit Materials*, ed. by A. Kelly, C. Zweben vol 3 Metal-Matrix Composits, ed. T.W. Clyne (Elsevier Amsterdam 2000) pp 321–340
41. Y. Yoshizawa, K. Yamauchi: Mater. Transact. JIM **31** 307 (1990)
42. T. Spassov, P. Solsona, S. Surinach, M.D. Baro: J. Alloys and Compounds **345** 123 (2002)
43. H. Gleiter: Nanometer-sized Materials In: *Encyclopedia of Physical Science and Technology*, 1991 Yearbook (Academic Press 1991) pp. 375–389

44. http://www.pvd-coatings.co.uk/theory-of-pvd-coatings-magnetron-sputtering.htm
45. H. Natter, R. Hempelmann: Electrochem. Acta **49** 51 (2003)
46. K. Tomantschger, G. Palumbo, F. Gonzales, H. Natter, R. Hempelmann, F. Endres, U. Erb, K.T. Aust: Electrochemical Synthesis of Nanocrystalline Materials In: *Jahrbuch Oberflchentechnik 2004*, ed. by R. Suchentrunk (Eugen G. Leuze Verlag, Bad Saulgau 2004) pp 23–43
47. A. Giga, Y. Kimoto, Y. Takigawa, K. Higashi: Scrip. Mater. **55** 143 (2006)
48. J. Rawers, D. Cook: NanoStruct.Mater. **11** 331 (1999)
49. Y.T. Feng, K. Han, D.R.J. Owen: Mater. Sci. Engineer. A **375–377** 815 (2004)
50. P. Pochet, P. Bellon, L. Chaffron, G. Martin: Mater. Sci. Forum **225–227** 207 (1996)
51. K. Suzuki, K. Sumiyama: Mater. Trans. JIM **36** 188 (1995)
52. Z. Wang, A.L. Fan, W.h. Tian, Y.T. Wang, X.G. Li: Mater. Lett.**60** 2227 (2006)
53. K. Lu, J.T. Wang, W.D. Wei: J. Appl. Phys. **69** 522 (1991). K. Lu: Mater. Sci. Engin. R **16** 161–221 (1996)
54. A.L. Greer : Nanocrystals Obtained from Devitrification. In: *Nanostructured Materials, Science and Technology* ed. by G.-M. Chow, N.I. Noskova (Cluver, Dordrecht 1998) pp 143–162
55. G. Wilde, N. Boucharat, R.J. Hebert, H. Rösner, W.S. Tong, J.H. Perepezko: Adv. Engin. Mater. **5** 125 (2003)
56. S. Surinach, E. Illekova, M.D. Baro, A. Jha, S. Jordery, M. Poulain, A. Soufiane, E.R. Taylor, D.N. Payne : Evaluation of Crystal Nucleation and Growth from Crystallization Kinetics Data of New Fluoride Glasses. In: *Nanostructured andNon-Crystalline Materials* ed. by M. Vazquez, A. Hernando (World Scientific,Singapore, New Jersey, London, Hong Kong 1995) pp 352–356
57. U. Köster, R. Janlewing: Mater. Sci. Engineer. **375-377** 223 (2004)
58. D. Jacovkis, J. Rodriguez-Viejo, M.T. Clavaguera-Mora: J. Phys.: Condens. Matter **17** 4897 (2005)
59. Y.B. Wang, H.W. Yang, B.B. Sun, B. Wu, J.Q. Wang, M.L. Sui, E. Ma: Scripta Mater. **55** 469 (2006)
60. I. Bokonyi, A. Cziraki: Nanostr. Mater. **11** 9 (1999)
61. H. Van Swygenhoven: Science **296** 66 (2002)
62. G.Z. Voronoi: J. Reine Angew. Math. **134** 199 (1908)
63. S. Enzo: Mater. Sci. Forum **269–272** 363(1998)
64. C.E. Krill, R. Haberkorn, R. Birringer: Specification of Microstructures and Characterization by Scattering techniques. In: *Handbook of Nanostructured Materials and Nanotechnology*, ed. by H.S. Nalwa (Academic Press, London, San Diego 2000) pp 155–212
65. L.S. Birks, M. Friedman: J. Appl. Phys. **17** 686 (1946)
66. B.E. Warren, B.L. Averbach: J. Appl. Phys. **23** 1059 (1952)
67. P. Chieux: Liquid Structure Investigation by Neutron Scattering In: *Neutron Diffraction*, Topics in Current Physics, vol 6, ed. by H. Dachs (Springer, Berlin Heidelberg 1978) pp 271–302
68. M. Lentzen, B. Jahnen, C.L. Jia, A. Thust, R.G.E. Tillmann, K. Urban: Ultramicroscopy **92** 233 (2002)
69. C.L. Jia, M. Lentzen, K. Urban: Microscopy and Microanal. **10** 174 (2004)
70. R. Tillmann, A. Thust, A. Gerber, M. Weides, K. Urban: Microscopy and Microanal. **11** 534 (2005)

71. S. Mentese, F. Juranyi, M. Scheffer, L.T. Hung, J.-B. Suck, G. Cuello in *Sci.Metastab.and Nanocryst.Alloys, Structure,Properties and Modelling* ed. by A.R. Dinesen, M. Eldrup, D. Juul Jensen, S. Linderoth, T.B. Pedersen, N.H. Pryds, A. Schrøder Pedersen, J.A. Werth (Risø Nat.Lab., Roskilde 2001) p. 329
72. H. Frase, L.J. Nagel, J.L. Robertson, B. Fultz: Phil. Mag B **75** 335 (1997)
73. T. Haubold, R. Birringer, B. Lengeler, H. Gleiter: Phys. Lett. A **135** 461 (1989)
74. G. Le Caer, P. Delcroix, J. Foct: Mater. Sci. Forum **269–272** 409 (1998)
75. S. Trapp, C.T. Limbach, U. Gonser, S.J. Campbell, H. Gleiter: Phys. Rev. Lett. **75** 3760 (1995)
76. C.J. Chen: *Introduction to Scanning Tunneling Microscopy,*(Oxford University Press, New York 1993)
77. R. Wiesendanger: *Scanning Probe Microscopy and Spectroscopy,*(Cambridge University Press 1994)
78. Z. Osvath, G. Vertesi, L. Tapaszto, F. Weber, Z.E. Horvath, J. Gyulai, L.P. Biro: Mater. Sci. Engineer. C **26** 1194 (2006)
79. http://www.eng.uc.edu/ gbeaucag/Class/Analysis/Chapt8.html
80. *Modern Aspects of Small Angle Scattering*, NATO ASI Series C: Mathematical and Physical Sciences, vol 451,ed. by H. Brumberger(Kluver Academic Publishers, Dordrecht, Boston, London 1995)
81. http://www.isis.rl.ac.uk/largescale/loq/documents/sans.htm
82. C. Williams, R.P. May, A. Guinier: Small Angle Scattering of X-Rays and Neutrons, In: *Characterization of Materials*, Material Science and Technology, vol 28, ed. by E. Lifshin (VCH Verlags Gesellschaft Weinheim 1994) pp 611–656
83. M.I. Sui, K. Lu, W. Deng, L.Y. Xiong, S. Patu, Y.Z. He: Phys. Rev. B **44** 6466 (1991)
84. Z.F. Dong, K. Lu, I. Bokonyi: Nanaostruct. Mater. **11** 187 (1999)
85. U. Herr: Adv. Engin. Mater. **3** 889 (2001)
86. B. Fultz, H. Kuwano, H. Ouyang: J. Appl. Phys. **77** 3458 (1995)
87. A. Caro, H. Van Swygenhoven: Phys. Rev. B **63** 134101 (2001)
88. H. Van Swygenhoven, D. Farkas, A. Caro: Phys. Rev B **62** 831 (2000)
89. Y.T. Zhu, J.Y. Huang, J. Gubicza, T. Ungar, Y.M. Wang, E. Ma et al.: J. Mat. Research **18** 1908 (2003)
90. J.A. Eastman, M.R. Fitzsimmons: J. Appl. Phys. **77** 522 (1995)
91. R.Z. Valiev, E.V. Kozlov, Y.F. Yvanov, J. Lian, A.A. Nazarov, B. Baudelet: Acta Metall. Mater. **42** 2467 (1994)
92. S. Veprek, Z. Iqbal, H.R. Oswald, A.P. Webb: J. Phys. C **14** 295 (1981)
93. D. Wolf, J. Wang, S.R. Phillpot, H. Gleiter: Phys. Lett. A **205** 274 (1995)
94. S.-N. Luo, L. Zheng, O. Tschauner: Solid State Comm. **136** 71 (2005)
95. T.D. Shen, C.C. Koch, T.L. McCormick, R.J. Nemanich, J.Y. Huang, J.G. Huang: J. Mat. Res. **10** 139 (1995)
96. M. Chauhan, F.A. Mohamed: Mater. Sci. Engineer. A **427** 7 (2006)
97. K. Lu, R. Lück, B. Predel: Scrip. Metall. Mater. **28** 1387 (1993)
98. K. Lu, R. Lück, B. Predel: Mater. Sci. Engin. A **179** 536 (1994)
99. H.J. Fecht: Phys. Rev. Lett. **65** 610 (1990)
100. M. Wagner: Acta Metall. Mater. **40** 957 (1992)
101. A.J. Haslam, S.R. Phillpot, D. Wolf, D. Moldovan, H. Gleiter: Mater. Sci. Engin. A **318** 293 (2001)

102. D. Moldovan, V. Yamakov, D. Wolf, S.R. Phillpot: Phys. Rev. Lett. **89** 206101 (2002)
103. J. Chadwick: J. Phys.: Condensed Matt. **11** 129 (1999)
104. Ph. Buffat, J.-P. Borel: Phys. Rev. A **13**, 2287 (1976)
105. A.N. Goldstein, C.M. Echer, A.P. Alivisatos: Science **256** 1425 (1992)
106. K.K. Nanda, S.N. Sahu, S.N. Behera: Phys. Rev. A **66** 013208 (2002)
107. W.H. Qi: Physica B **368** 46 (2005)
108. W.H. Qi, M.P. Wang, M. Zhou, X.Q. Shen, X.F. Zhang: J. Phys. Chem. Solids **67** 851 (2006)
109. M.E. McHenry, D.E. Laughlin: Acta mater.**48** 223 (2000)
110. A.K. Menon, B.K. Gupta: Nanostruc. Mater. **12** 1117 (1999)
111. A. Hernando: europhys. news **34** 232 (2003)
112. K. Twarowski, M. Kuzminski, A. Slawska-Waniewska, H.K. Lachowicz, G. Herzer: On the Origin of Effective Linear Magnetostriction in FeCuNbSiB Nanocrystalline Alloys. In: *Nanostructured and Non-Crystalline Materials*, ed. by M. Vazquez, A. Hernando (World Scientific, Singapore, New Jersey, London, Hong Kong 1995) pp 495
113. G. Herzer: J. Mag. Mag. Mater. **157–158** 133 (1996)
114. G. Herzer: Nanocrystalline Soft Magnetic Alloys. In: *Handbook of Magnetic Materials* Vol. **10** ed. by K.H.J. Buschow (Elsevier, Amsterdam 1997) pp 415–462
115. G. Herzer: IEEE Transact. Magnet. **25** 3327 (1989)
116. A. Slawska-Waniewska, M. Grafoute, J.M. Greneche: J. Phys.: Condensed Matt. **18** 2235 (2006)
117. G. Herzer: Mater. Sci. Engineer A **133** 1 (1991)
118. A. Bahrami, H.R. Madaah Hosseini, P. Abachi, S. Miraghaei: Mater. Lett. **60** 1068 (2006)
119. R. Alben, J.J. Becker, M.C. Chi: J. Appl. Phys. **49** 1653 (1978)
120. J.F. Löffler, H.-B. Braun, W. Wagner: Phys. Rev. Lett. **85** 1990 (2000)
121. J.F. Löffler, W. Wagner, G. Kostorz: J. Appl. Cryst. **33** 451 (2000)
122. J.R. Weertman: *Mechanical Behaviour of Nanocrystalline Metals in Nanostructured Materials: Processing, Properties and Potential Applications*, (William Andrew, Norwich, New York 2002)
123. V. Yamakov, D. Wolf, S.R. Phillpot, A.K. Mukherjee, H. Gleiter: nature materials **1** 45 (2002)
124. K.W. Jacobsen, J. Schiotz: nature materials **1** 15 (2002)
125. Z. Budrovic, H. Van Swygenhoven, P.M. Derlet, S. Van Petegem, B. Schmitt: Science **304** 273 (2004)
126. H. Van Swygenhoven, P.M. Derlet, A. Hasuaoui: Phys. Rev. B **66** 024101 (2002)
127. J. Schiotz, K.W. Jacobsen: Science **301** 1357 (2003)
128. E.O. Hall: Proc. Phys. Soc. London B **64** 747 (1951)
129. N.J. Petch: J. Iron Steel Inst. **174** 25 (1953)
130. V. Yamakov, D. Wolf, S.R. Phillpot, A.K. Mukherjee, H. Gleiter: nature materials **3** 43 (2004)
131. J. Chen, L. Lu, K. Lu: Script. Mater. **54** 1913 (2006)
132. S. Yip: Nature **391** 532 (1998)
133. J. Schiotz, F.D. Di Tolla, K.W. Jacobsen: Nature **391** 561 (1998)
134. S. Yip: nature materials **3** 11 (2004)

135. H. Van Swygenhoven, P.M. Derlet, A.G. Froseth: nature materials **3** 399 (2004)
136. V. Yamakov, D. Wolf, S.R. Phillpot, A.K. Mukherjee, H. Gleiter: Phil. Mag. Lett. **83** 385 (2003)
137. H. Van Swygenhoven, M. Spaczer, A. Caro: Acta Mater. **47** 3117 (1999)
138. H. Van Swygenhoven, P.M. Derlet: Phys. Rev. B **64** 224105 (2001)
139. R.L.A. Coble: J. App. Phys. **34** 1679 (1963)
140. H. Van Swygenhoven, Z. Budrovic, P.M. Derlet, A.G. Froseth, S. Van Petegem: Mat. Sci. Engin. A **400–401** 329 (2005)
141. M. Chen, E. Ma, K.J. Hemker, H. Sheng, Y. Wang, X. Cheng: Science **300** 1275 (2003)
142. Z. Shan, E.A. Stach, J.M.K. Wiezorek, J.A. Knapp, D.M. Follstaedt, S.X. Mao: Science **305** 654 (2004)
143. E. Ma: Science **305** 623 (2004)
144. Y.M. Wang, E. Ma: Mater. Sci. Engineer. **375-377** 46 (2004)
145. Y.M. Wang,A.V. Hamza, E. Ma: Acta Mater. **54** 2715 (2006)
146. B. Zhu, R.S. Asaro, P. Krysl, k. Zhang, J.R. Weertman: Acta Mater. **54** 3307 (2006)
147. Z. Iqbal, S. Veprek: J. Phys. C: Solid State Phys. **15** 377 (1982)
148. P. Verma, L. Gupta, S.C. Abbi, K.P. Jain: J. App. Phys. **88** 4109 (2000)
149. J.E. Spanier, R.D. Robinson, F. Zhang, S.W. Chan, I.P. Herman: Phys. Rev. B **64** 245047 (2001)
150. C.H. Shek, G.M. Lin, J.K.L. Lai: Nanostruc. Mater. **11** 831 (1999)
151. M.J. Seong, O.I. Micic, A.J. Nozik, A. Mascarenhas, H.M. Cheong: Appl. Phys. Lett. **82** 185 (2003)
152. M. Gotic, M. Ivanda, A. Sekulic, S. Music, S. Popovic, A. Turkovic, K. Furic: Mater. Lett **28** 225 (1996)
153. J.-B. Suck, H. Rudin, H.-J. Güntherodt, H. Beck: J. Phys. C: Solid State Phys. **14** 2305 (1981)
154. P.M. Derlet, R. Meyer, L.J. Lewis, U. Stuhr, H. Van Swygenhoven: Phys. Rev. Lett. **87** 205501 (2001)
155. A.A. Maradudin, E.W. Montroll, G.H. Weiss in *Solid State Phys.* Suppl. 3 ed. by F. Seiz, D. Turnbull (Academ. Press NewYork, London 1963)
156. A.A. Maradudin, E.W. Montroll, G.H. Weiss, I.P. Ipatova in *Solid State Phys.* (2nd edit.) ed. by F. Seiz, D. Turnbull (Academ. Press New York, London 1971)
157. R. Singh, S. Prakash: Chinese J. Phys. **40** 624 (2002)
158. B. Fultz, J.L. Robertson, T.A. Stephens, L.J. Nagel, S. Spooner: J. App. Phys. **79** 8318 (1996)
159. B. Fultz, C.C. Ahn, E.E. Alp, W. Sturhahn, T.S. Toellner: Phys. Rev. Lett. **79** 937 (1997)
160. H. Frase, B. Fultz, J.L. Robertson: Phys. Rev. B **57** 898 (1998)
161. E. Bonetti, L. Pasquini, E. Sampaolesi, A. Deriu, G. Cicognani: J. Appl. Phys. **88** 4571 (2000)
162. L. Pasquini, A. Barla, A.I. Chumakov, O. Leupold, R. Rüffer, A. Deriu, E. Bonetti: Phys. Rev. B **66** 073410 (2002)
163. S. Mentese, J.-B. Suck, S. Janssen: Physica B **316-317** 438 (2002)
164. D. Wolf, J. Wang, S.R. Phillpot, H. Gleiter: Phys. Rev. Lett. **74** 4686 (1995)
165. J. Wang, D. Wolf, S.R. Phillpot, H. Gleiter: Phil. Mag A **73** 517 (1996)
166. A. Kara, T.S. Rahman: Phys. Rev. Lett. **81** 1453 (1998)

167. S. Mentese, J.-B. Suck, O.A. Petrenko unpublished
168. S. Mentese, J-B. Suck, V. Reat: Appl. Phys. A **74** S969 (2002)
169. J.-B. Suck, H. Bretscher, H. Rudin, P. Grütter, H.-J. Güntherodt: Phys. Rev. Lett. **59** 102 (1987)
170. J.-B. Suck, H.-J. Güntherodt: Icosahedral, Glassy and Crystallise $Al_{75}Cu_{15}V_{10}$: A Comparative Study of their Generalised Vibrational Density-of-States. In: *Phonons 89*, ed. by S. Hunklinger, W. Ludwig, G. Weiss (World Scientific, Singapore 1990) p. 573
171. J.-B. Suck: Vibrational Density-of-States of Metastable and Stable Quasicrystalline Alloys. In: *Quasicrystals*, ed. by J.-B. Suck, M. Schreiber, P. Häussler (Springer, Berlin Heidelberg New York 2002) pp 454
172. D. Korn, A. Morsch, R. Birringer, W. Arnold, H. Gleiter: J. Physique (Paris) **49** C5-769 (1988)
173. D.Y. Sun, X.G. Gong, X.-Q. Wang: Phys. Rev. B **63** 193412 (2001)
174. A. Tamura, K. Higeta, T. Ichinokawa: J. Phys. C: Solid State **15** 4975 (1982)
175. A. Tamura, K. Higeta, T. Ichinokawa: J. Phys. C: Solid State **16** 1585 (1983)
176. A. Tamura, T. Ichinokawa: J. Phys. C: Solid State **16** 4779 (1983)
177. J.J. Burton: J. Chem. Phys. **52** 345 (1970)
178. J.-B. Suck: J. Non-crystall. Solids **153-154** 573 (1993)
179. S. Mentese, J.-B. Suck, A.J. Dianoux: J. Metastab. and Nanocryst. Materials **8** 671 (2000)
180. K. Lu, R. Lück, B. Predel: J. Non-Cryst. Solids **156–158** 589 (1993)
181. J. Trampenau, K. Bauszus, W. Petry, U. Herr: Nanostr. Mater. **6** 551 (1995)
182. U. Stuhr, H. Wipf, K.H. Andersen, H. Hahn: Phys. Rev. Lett. **81** 1449 (1998)
183. U. Stuhr, H. Wipf, T.J. Udovic, J. Weißmüller, H. Gleiter: Nanostruc. Mater. **6** 555 (1995)
184. J.-B. Suck, H. Rudin: Vibrational Dynamics of Metallic Glasses Studied by Neutron Inelastic Scattering. In: *Glassy Metals II*, Topics in Applied Physics **53** ed. by H. Beck, H.-J. Güntherodt (Springer, Berlin Heidelberg New York 1983) pp 217
185. S. Mentese, thesis, University of Technology Chemnitz, Germany 2002
186. J.-B. Suck, S. Mentese, S. Jannsen submitted

Index

Σ-value   94
α-quartz   14
β-quartz   14
2D Bravais lattices   34

AlN tubes   47
Andersen barostat   9
anisotropy energy   162
application of nanomaterials   148
armchair nanotubes   37
Arrhenius plot   27
atomic force microscopy   157
atomic shuffling   166
atomistic modelling   99
   Born model   103
   EAM   102
   shell model   104

B nanotubes   43
band structure   64
Bi nanotubes   43
Bloch's theorem   100
BN nanotubes   45
Boltzmann theory   68
boride nanotubes   45
boundaries   92, 97
boundary
   grain   92
   heterophase   92, 97, 109
   homophase   92, 104
   interaction terms   95
   tilt   105
   twist   105
Brillouin zone   64

bulk viscosity   11

Car Parrinello Molecular Dynamics   3
CaSi$_2$ nanotubes   42
chalcogenide nanotubes   48
chemical potential   160
chemical potentials   108
chiral nanotubes   37
chiral vector   35
coercivity   164
cohesive energy   160
coincidence site lattice   94
complex wave vector   81
CSL   105
current-current correlation functions
   26
current-voltage characteristics   84

damped harmonic oscillator   171
dangling bonds   29
Debye cut-off frequency   170
Debye temperature   20, 170
Debye–Waller coefficient   170
density functional theory   40
   APW   100
   MB-PP   100
   PW-PP   100
density functional theory (DFT)   3
density of states   74
density-functional theory   62, 99
detailed balance   2
difference GVDOS   181
Differential Scanning Calorimetry   157
diffusion equation   9

diffusive transport   67
dislocation pile up   165
Dulong–Petit law   24
dynamic structure factor   168

effective coupling constant   163
effective exchange stiffness   164
Einstein relation   10
electrodeposition   153
electron gas parameter   98
electronegativity   97
electrostatic energy
    nanotube   40
embedded atom model   102
entropy   170
ergodicity hypothesis   2
Ewald summation   12
exchange correlation length   162
exchange energy   162
Extended X-ray Absorption Fine
    Structure   156

Fermi surface   64
Fermi velocities   64
FINEMET   162
finite life time   171
fluctuating charge potential   21
fluctuating dipole moment potential
    21
fourth-order cumulant   18
Frank–Read sources   164
free volume   157, 159
Friedel oscillations   65

GaN tubes   47
GB-diffusion   166
GB-sliding   166
Ge nanotubes   41
generalized random anisotropy model
    163
generalized vibrational density-of-states
    169
giant magnetoresistance   60, 72
GMR ratio   72
grain boundary energy   106
grain growth   173
grain-boundary   149
grand canonical potential   108
Green–Kubo relations   11

growth   110
    Franck-van-der-Merwe   111
    Stranski–Krastanov   96, 112
    Volmer–Weber   112
Guinnier plots   157

Hall–Petch relation   165
Helmholtz free energy   170
hydrodynamic slowing down   10
hydrogen plasma-metal reaction   153

image charges   98, 101
inelastic structure factor   181
inorganic nanotubes   33
instability of nc material   159
interface   91, 173
    coherent   94
    commensurate   94
    degrees of freedom   92, 93
    incommensurate   94
    reactive   116
interface-modes   179
intergrain region   149
interlayer exchange coupling   60
internal energy   170

Julliere model   79

KKR method   63

Landau theory   18
Landauer-Büttiker theory   70
linear response   69
Liouville operator   6
Liouville's theorem   6
log-normal-distribution   151

Mössbauer spectroscopy   156
MA production parameters   153
magnetic anisotropy   162
magnetic field sensor   78
magnetic properties   161
magnetic random access memory   85
magnetic tunnel junction   60
magnetostriction   162
Maxwell–Boltzmann distribution   7, 8
mean force constant   170
mean free path   171
mean square vibrational amplitude
    171
mechanical attrition   153

melting temperature of nanocrystals
  160
metallic multilayer   64
microscopic transition probability   68
misfit   95
mixing rule   150
mode coupling theory   27
molecular dynamics (MD)   1
molecular dynamics simulations   1
Monte Carlo (MC)   2
Moore's law   59
MoS$_2$ tubes   47

nano-crystallization   154
nano-materials   147
nano-quenching   154
nano-technologies   148
nanocable   45
nanocrystalline quasicrystals   173
nanoroll   51
nanostrip   46, 51
nanotube caps   37
nanotubes
  classification   35
  doping   50, 51
  group IVA elements   41
  group VA elements   43
  physical properties   37
Newton's equations of motion   2, 4, 8
nitride nanotubes   47
Nosé–Hoover algorithm   8

oxide nanostructures   50

P nanotubes   43
Parrinello–Rahman method   15
partial charge   96
partial density-of-states   169
partial dynamic structure factors   169
path integral molecular dynamics   4,
  20
path integral Monte Carlo (PIMC)   4
Pauli repulsion   98
permeability   164
phonon confinement   171
physical length scales   148
plasticity map   165
polycrystals   148
positron-lifetime measurements   157

pressure tensor   11

quantum confinement   75
quantum well states   75

radial distribution function   156
Raman scattering   168
random anisotropy model   163
reactive interlayer   115
reading heads   78
relaxation time approximation   69
residual dislocation network   167
residual resistivity   69
reverse Hall–Petch mechanism   165
rigid band model   51
RKKY-like interaction   65

saturation magnetization   163
scanning probe microscopy   157
second-order phase transition   19
selenide tubes   50
self diffusion constants   27
semiempirical methods   40
shear viscosity   11
short range order   156
Si nanotubes   41
specific conductivity   67
specific heat   23, 170
sputtering   152
stacking fault energy   165
static structure factor   10, 156
Stoner model   61
strain energy
  nanotube   38, 40
structural relaxation   173, 178
structure factor   28
sulfide tubes   50
surface modes   180
surface tensions   111

thermal conductivity   12
thermodynamic properties   170
thermostats   3
tight-binding   100
tight-binding method   50
total dynamic structure factor   169
transmission electron microscopy   156
transmission probability   82
tricluster   1

tunneling magnetoresistance   60, 78
two-current-model   70
two-level systems   24

undercooled fluids and glasses   1

vanadate nanotubes   53
vapour phase production   151
vector mean free path   69
Verlet algorithm   5, 6
vibrational density-of-states   168
virial theorem   7

wetting   110

angle   111
wide angle X-ray and neutron diffraction
    155
work of separation   106
WS_2 tubes   47

X-ray and Neutron Small Angle
    Scattering   157
X-ray diffraction   15

Young's equation   111

zigzag nanotubes   37